LECTURES IN NONLINEAR MECHANICS AND CHAOS THEORY

LECTURES IN NONLINEAR MECHANICS AND CHAOS THEORY

Albert W. Stetz

Oregon State University, USA

 World Scientific

NEW JERSEY · LONDON · SINGAPORE · BEIJING · SHANGHAI · HONG KONG · TAIPEI · CHENNAI · TOKYO

Published by

World Scientific Publishing Co. Pte. Ltd.

5 Toh Tuck Link, Singapore 596224

USA office: 27 Warren Street, Suite 401-402, Hackensack, NJ 07601

UK office: 57 Shelton Street, Covent Garden, London WC2H 9HE

Library of Congress Cataloging-in-Publication Data

Names: Stetz, Albert W., 1940–   author.

Title: Lectures on nonlinear mechanics and chaos theory / Albert W. Stetz,
    Oregon State University, USA.

Description: Singapore ; Hackensack, NJ : World Scientific Publishing Co. Pte. Ltd., [2016] | 2016 |
    Includes bibliographical references and index.

Identifiers: LCCN 2016008403| ISBN 9789813141353 (hardcover ; alk. paper) |
    ISBN 9813141352 (hardcover ; alk. paper)

Subjects: LCSH: Nonlinear mechanics. | Nonlinear systems. | Chaotic behavior in systems.

Classification: LCC QA805 .S724 2016 | DDC 531--dc23

LC record available at http://lccn.loc.gov/2016008403

British Library Cataloguing-in-Publication Data

A catalogue record for this book is available from the British Library.

In-house Editor: Christopher Teo

Typeset by Stallion Press
Email: enquiries@stallionpress.com

Printed in Singapore

Preface

This book is based on a one-quarter course in graduate mechanics that has been given in the Physics Department of Oregon State University in recent years. The students who take it are typically first-year graduate students who have had an upper-division undergraduate course in mechanics where they were introduced to the Lagrangian and Hamiltonian formulations. We review this material and then discuss more advanced subjects such as transformation theory and perturbation theory. Finally, we give our students an introduction to modern chaos theory including the theory of maps, Lyapunov exponents, the Poincaré-Birkhoff theorem, and the KAM theorem. The material in this book fits nicely into the ten-week quarter format of our classes. It also contains problems to illustrate the material in each chapter and an assortment of more advanced projects requiring some expertise in computing.

I wish to thank my students who over the years have taught me what works and what doesn't work in the classroom. I would also like to give my thanks to our chairman Henri Jansen for making all this possible, to my illustrators Rachel Dallas and Robin Coleman, and especially to my daughter Karelia Stetz-Waters who with her expertise as English professor and novelist has coached me in the business of publishing.

Contents

Chapter 1

Lagrangian dynamics

1.1 Introduction

The possibility that deterministic mechanical systems could exhibit the behavior we now call chaos was first realized by the French mathematician Henri Poincaré sometime toward the end of the nineteenth century. His discovery emerged from analytic or classical mechanics, which is still part of the foundation of physics. To oversimplify a bit, classical mechanics deals with those problems that can be "solved," in the sense that it is possible to derive equations of motions that describe the positions of the various parts of a system as functions of time using standard analytic functions. Nonlinear dynamics treats problems that cannot be so solved, and it is only in these problems that chaos can appear. The simple pendulum makes a good example. The differential equation of motion is

$$\ddot{\theta} + \omega^2 \sin\theta = 0 \tag{1.1}$$

The $\sin\theta$ is a nonlinear function of θ. If we linearize by setting $\sin\theta \approx \theta$, the solutions are elementary functions, $\sin\omega t$ and $\cos\omega t$. If we keep the $\sin\theta$, the solutions can only be expressed in terms of elliptic integrals. This is not a chaotic system, because there is only one degree of freedom, but if we hang one pendulum from the end of another, the equations of motion are hopeless to find (even with elliptic integrals) and the resulting motion can be chaotic.[1]

In order to arrive at Poincaré's moment of discovery, we will have to review the development of classical mechanics through the nineteenth century. This material is found in many standard texts, but I will cover it here

---

[1]I should emphasize the distinction between the *differential* equations of motion, which are usually simple (though nonlinear), and the equations that describe the positions of the elements of the system as functions of time, which are usually non-existent except possibly in the form of an infinite series.

in some detail. This is partly to insure uniform notation throughout these lectures and partly to focus on those things that lead directly to chaos in nonlinear systems. We will begin formulating mechanics in terms of generalized coordinates and the Lagrange equations. We then study Legendre transformations and use them to derive Hamilton's equations of motion. These equations are particularly suited to conservative systems in which the Hamiltonian is constant in time, and it is such systems that will be our primary concern. It turns out that Legendre transformations can be used to transform Hamiltonians in a myriad of ways. One particularly elegant form uses action-angle variables to transform a certain class of problems into a set of uncoupled harmonic oscillators. Systems that can be so transformed are said to be integrable, which is to say that they can be "solved," at least in principle. What happens, Poincaré asked, to a system that is almost but not quite integrable? The answer entails perturbation theory and leads to the disastrous problem of small divisors. This is the path that led originally to the discovery of chaos, and it is the one we will pursue here.

1.2   Generalized coordinates and the Lagrangian

Vector equations, like $\boldsymbol{F} = m\boldsymbol{a}$, seem to imply a coordinate system. Beginning students learn to use cartesian coordinates and then learn that this is not always the best choice. If the system has cylindrical symmetry, for example, it is best to use cylindrical coordinates: it makes the problem easier. By "symmetry" we mean that the number of degrees of freedom of the system is less that the dimensionality of the space in which it is imbedded. The familiar example of the block sliding down the incline plane will make this clear. Let's say that it's a two dimensional problem with an x-y coordinate system. The block is constrained to move in a straight line, however, so that its position can be completely specified by one variable, *i.e.* it has one degree of freedom. The clever student chooses the x axis so that it lies along the path of the block. This reduces the problem to one dimension, since $y = 0$ and the x coordinate of the block is given by one simple equation. In the pendulum example from the previous section, it was most convenient to use a polar coordinate system centered at the pivot. Since r is constant, the motion can be described completely in terms of θ.

These coordinate systems conceal a subtle point: the pendulum moves in a circular arc and the block moves in a straight line because they are acted on by forces of constraint. In most cases we are not interested in these forces. Our choice of coordinates simply makes them disappear from the problem. Most problems don't have obvious symmetries, however. Consider a bead sliding along a wire following some complicated snaky path in three-dimensional space. There's only one degree of freedom, since the particle's position is determined entirely by its distance measured along the wire from some reference point. The forces are so complicated however, that it is out of the question to solve the problem by using $\boldsymbol{F} = m\boldsymbol{a}$ in any straightforward way. This is the problem that Lagrangian mechanics is designed to handle. The basic (and quite profound) idea is that even though there may be no *coordinate system* (in the usual sense) that will reduce the dimensionality of the problem, yet there is usually a *system of coordinates* that will do this. Such coordinates are called *generalized coordinates*. To be more specific, suppose that a system consists of N point masses with positions specified by ordinary three-dimensional cartesian vectors, $\boldsymbol{r}_i$, $i = 1 \cdots N$, subject to some constraints. The easiest constraints to deal with are those that can be expressed as a set of l equations of the form

$$f_j(\boldsymbol{r}_1, \boldsymbol{r}_2, \ldots, t) = 0 \qquad (1.2)$$

where $j = 1 \cdots l$. Such constraints are said to be *holonomic*. If in addition, the equations of constraint do not involve time explicitly, they are said to be *scleronomous*, otherwise they are called *rheonomous*. These constraints can be used to reduce the $3N$ cartesian components to a set of $3N - l$ variables $q_1, q_2, \ldots, q_{3N-l}$. The relationship between the two is given by a set of N equations of the form

$$\boldsymbol{r}_i = \boldsymbol{r}_i(q_1, q_2, \ldots, q_{3N-l}, t) \qquad (1.3)$$

The q's used in this way are the generalized coordinates. In the example of the bead on a curved wire, the equations would reduce to $\boldsymbol{r} = \boldsymbol{r}(q)$, where q is a distance measured along the wire. This simply specifies the curvature of the wire.

It should be noted that the q's need not all have the same units. Also note that we can use the same notation even if there are no constraints. For example, the position of an unconstrained particle could be written $\boldsymbol{r} = \boldsymbol{r}(q_1, q_2, q_3)$, and the q's might represent cartesian, spherical, or cylindrical coordinates. In order to simplify the notation, we will often pack the q's into an array and use vector notation,

$$\boldsymbol{q} = \begin{vmatrix} q_1 \\ q_2 \\ q_3 \\ \vdots \end{vmatrix} \qquad (1.4)$$

This is not meant to imply that $\boldsymbol{q}$ is a vector in the usual sense. For one thing, it does not necessarily posses "a magnitude and a direction" as good vectors are supposed to have. By the same token, we cannot use the notion of orthogonal unit vectors.

Along with the notion of generalized coordinates comes that of generalized velocities.

$$\dot{q}_k \equiv \frac{dq_k}{dt} \qquad (1.5)$$

Since q_i depends only on t, this is a total derivative, but when we differentiate $\boldsymbol{r}_i$, we must remember that it depends both explicitly on time as well as implicitly through the q's.

$$\dot{\boldsymbol{r}}_i = \sum_k \frac{\partial \boldsymbol{r}_i}{\partial q_k} \dot{q}_k + \frac{\partial \boldsymbol{r}_i}{\partial t} \qquad (1.6)$$

(In this chapter I will consistently use the index i to sum over the N point masses and k to sum over the $3N - l$ degrees of freedom.) Differentiating both sides with respect to $\dot{q}_k$ yields

$$\frac{\partial \dot{\boldsymbol{r}}_i}{\partial \dot{q}_k} = \frac{\partial \boldsymbol{r}_i}{\partial q_k} \qquad (1.7)$$

which will be useful in the following derivations.

1.3   Virtual work and generalized force

There are several routes for deriving Lagrange's equations of motion. The most elegant and general makes use of the principle of least action and the calculus of variation. I will use a much more pedestrian approach based on Newton's second law of motion. First note that $\boldsymbol{F} = m\boldsymbol{a}$ can be written in the rather arcane form

$$\frac{d}{dt} \left(\frac{\partial T}{\partial v_i} \right) = F_i \qquad (1.8)$$

where F_i is i-th component of the total force acting on a particle with kinetic energy T. The point of writing this in terms of energy rather than acceleration is that we can separate out the forces of constraint, which are always perpendicular to the direction of motion and hence do no work. The trick is to write this in terms of generalized coordinates and velocities. This is rather technical, but the underlying idea is simple, and the result looks much like (1.8).

The q_k's are all independent, so we can vary one by a small amount δq_k while holding all others constant.

$$\delta \boldsymbol{r}_i = \sum_k \frac{\partial \boldsymbol{r}_i}{\partial q_k} \delta q_k \tag{1.9}$$

This is sometimes called a virtual displacement. The corresponding virtual work is

$$\delta W_k = \sum_i \boldsymbol{F}_i \cdot \left(\frac{\partial \boldsymbol{r}_i}{\partial q_k} \delta q_k \right) \tag{1.10}$$

We define a generalized force

$$\Im_k \equiv \sum_i \boldsymbol{F}_i \cdot \frac{\partial \boldsymbol{r}_i}{\partial q_k} \tag{1.11}$$

The forces of constraint can be excluded from the sum for the reason explained above. We are left with

$$\Im_k = \frac{\delta W_k}{\delta q_k} \tag{1.12}$$

The kinetic energy is calculated using ordinary velocities.

$$T = \frac{1}{2} \sum_i m_i \, \dot{\boldsymbol{r}}_i \cdot \dot{\boldsymbol{r}}_i \tag{1.13}$$

$$\frac{\partial T}{\partial q_k} = \sum_i m_i \, \dot{\boldsymbol{r}}_i \cdot \frac{\partial \dot{\boldsymbol{r}}_i}{\partial q_k} = \sum_i \boldsymbol{p}_i \cdot \frac{\partial \dot{\boldsymbol{r}}_i}{\partial q_k} \tag{1.14}$$

$$\frac{\partial T}{\partial \dot{q}_k} = \sum_i m_i \, \dot{\boldsymbol{r}}_i \cdot \frac{\partial \dot{\boldsymbol{r}}_i}{\partial \dot{q}_k} = \sum_i \boldsymbol{p}_i \cdot \frac{\partial \boldsymbol{r}_i}{\partial q_k} \tag{1.15}$$

Equation (1.7) was used to obtain the last term. A straightforward calculation now leads to

$$\Im_k = \frac{d}{dt} \left(\frac{\partial T}{\partial \dot{q}_k} \right) - \frac{\partial T}{\partial q_k} \tag{1.16}$$

which is the generalized form of (1.8).

1.4   Conservative forces and the Lagrangian

So far we have made no assumptions about the nature of the forces included in $\Im$ except that they are not forces of constraint.   Equation (1.16) is therefore quite general, although seldom used in this form. In these notes we are primarily concerned with conservative forces, *i.e.* forces that can be derived from a potential.

$$\boldsymbol{F}_i = -\boldsymbol{\nabla}_i V(\boldsymbol{r}_1 \cdots \boldsymbol{r}_N) \tag{1.17}$$

Notice that V doesn't depend on velocity. (Electromagnetic forces are velocity dependent of course, but they can easily be accommodated into the Lagrangian framework. I will return to this issue later on.) Now calculate the work done by changing some of the q's.

$$
\begin{aligned}
W &= \int \sum_i \boldsymbol{F}_i \cdot d\boldsymbol{r}_i = -\sum_i \int \boldsymbol{\nabla}_i V \cdot d\boldsymbol{r}_i \\
&= -\sum_i \int \boldsymbol{\nabla}_i V \cdot \sum_k \frac{\partial \boldsymbol{r}_i}{\partial q_k} dq_k \\
&= -\sum_k \int \left(\sum_i \boldsymbol{\nabla}_i V \cdot \frac{\partial \boldsymbol{r}_i}{\partial q_k} \right) dq_k \\
&= -\sum_k \int \frac{\partial V}{\partial q_k} dq_k
\end{aligned}
\tag{1.18}
$$

The integral is a multidimensional definite integral over the various q's that have changed. We could also calculate W by summing and integrating (1.12).

$$\delta W = \sum_k \delta W_k = \sum_k \Im_k \delta q_k \tag{1.19}$$

$$W = \sum_k \int \Im_k dq_k \tag{1.20}$$

Comparison with (1.18) yields

$$\Im_k = -\frac{\partial V}{\partial q_k} \tag{1.21}$$

Finally define the Lagrangian

$$L = T - V \tag{1.22}$$

Equation (1.16) becomes

$$\frac{d}{dt}\left(\frac{\partial L}{\partial \dot{q}_k}\right) - \frac{\partial L}{\partial q_k} = 0 \tag{1.23}$$

Equation (1.23) represents a set of $3N - l$ second-order partial differential equations called Lagrange's equations of motion. I can summarize this long development by giving you a "cookbook" procedure for using (1.23) to solve mechanics problems: First select a convenient set of generalized coordinates. Then calculate T and V in the usual way using the r_i's. Use equation (1.3) to eliminate the r_i's in favor of the q_k's. Finally substitute L into (1.23) and solve the resulting equations.

Classical mechanics texts are full of examples in which this program is carried to a successful conclusion. In fact, most of these problems are contrived and of little interest except to illustrate the method. The vast majority of systems lead to differential equations that cannot be solved in closed form. The modern emphasis is to understand the solutions qualitatively and then obtain numerical solutions using the computer. The Hamiltonian formalism described in the next section is better suited to both these ends.

1.4.1 *The central force problem in a plane*

Consider the central force problem as an example of this technique.

$$V = V(r) \qquad \boldsymbol{F} = -\boldsymbol{\nabla} V \tag{1.24}$$

$$L = T - V = \frac{1}{2}m\left(\dot{r}^2 + r^2\dot{\phi}^2\right) - V(r) \tag{1.25}$$

Let's choose our generalized coordinates to be $q_1 = r$ and $q_2 = \phi$. Equation (1.23) becomes

$$m\ddot{r} - mr\dot{\phi}^2 + \frac{dV}{dr} = 0 \tag{1.26}$$

$$\frac{d}{dt}\left(mr^2\dot{\phi}\right) = 0 \tag{1.27}$$

This last equation tells us that there is a quantity $mr^2\dot{\phi}$ that does not change with time. Such a quantity is said to be conserved. In this case we have rediscovered the conservation of angular momentum.

$$mr^2\dot{\phi} \equiv l_z = \text{constant} \tag{1.28}$$

This reduces the problem to one dimension.

$$m\ddot{r} = \frac{l_z^2}{mr^3} - \frac{dV}{dr} \tag{1.29}$$

Since there are no constraints, the generalized forces are identical with the
ordinary forces

$$\Im_\phi = -\frac{dV}{d\phi} = 0 \qquad \Im_r = -\frac{dV}{dr} \qquad (1.30)$$

This equation has an elegant closed form solution in the special case of
gravitational attraction.

$$V = -\frac{GmM}{r} \equiv -\frac{k}{r} \qquad (1.31)$$

$$m\ddot{r} = \frac{l_z^2}{mr^3} - \frac{k}{r^2} \qquad (1.32)$$

This apparently nonlinear equation yields to a simple trick, let $u = 1/r$.

$$\frac{d^2u}{d\phi^2} + u = \frac{mk}{l_z^2} \qquad (1.33)$$

If the motion is circular u is constant. Otherwise it oscillates around the
value mk/l_z^2 with simple harmonic motion.[2] The period of oscillation is
identical with the period of rotation so the corresponding orbit is an ellipse.

This problem was easy to solve because we were able to discover a non-
trivial quantity that was constant, in this case the angular momentum. The
constant enabled us to reduce the number of independent variables from
two to one. Such a conserved quantity is called an *integral of the motion*
or a *constant of the motion*. Obviously, the more such quantities one can
find, the easier the problem. This raises two practical problems. First, how
can we tell, perhaps from looking at the physics of a problem, how many
independent conserved quantities there are? Second, how are we to find
them?

In the central force problem, both of these questions answered themselves. We know that angular momentum is conserved. This fact manifests
itself in the Lagrangian in that L depends on $\dot{\phi}$ but not on ϕ. Such a
coordinate is said to be *cyclic* or *ignorable*. Let q be such a coordinate.
Then

$$\frac{d}{dt}\left(\frac{\partial L}{\partial \dot{q}}\right) = 0 \qquad (1.34)$$

The quantity in brackets has a special significance. It is called the *canonically conjugate momentum*.[3]

$$\frac{\partial L}{\dot{q}_k} \equiv p_k \qquad (1.35)$$

---

[2]This illustrates a general principle in physics: When correctly viewed, everything is a
harmonic oscillator.

[3]This notation is universally used, hence the old aphorism that mechanics is a matter
of minding your p's and q's.

To summarize, if q is cyclic, p is conserved.

Suppose we had tried to do the central force problem in cartesian coordinates. Both x and y would appear in the Lagrangian, and neither p_x nor p_y would be constant. If we insisted on this, central forces would remain an intractable problem in two dimensions. We need to choose our generalized coordinates so that there are as many cyclic variables as possible. The two questions reemerge: how many are we entitled to and how do we find the corresponding p's and q's?

1.5 Noether's theorem

A partial answer to the first question is given by a well-known result called Noether's theorem: for every transformation that leaves the Lagrangian invariant there is a constant of the motion. This theorem (which underlies all of modern particle physics) says that there is a fundamental connection between symmetries and invariance principles on one hand and conservation laws on the other. Momentum is conserved because the laws of physics are invariant under translation. Angular momentum is conserved because the laws of physics are invariant under rotation.

The proof of the theorem is all its generality requires some heavy machinery, but the following simple proof will do for our purposes. For brevity I will sometimes use the notation q to refer to the array $(q_1, q_2, ...q_n)$. Let $q(t)$ be a solution of the equations of motion generated by the Lagrangian $L(q, \dot{q}, t)$. Suppose there is some transformation characterized by the parameter ϵ. Put another way, there is a function (let your imagination be your guide) $\varphi(t, \epsilon)$ such that

$$\varphi(t, 0) = q(t), \qquad \frac{\partial \varphi}{\partial t} = \dot{\varphi}, \qquad \frac{\partial \varphi_i}{\partial \epsilon} \equiv \delta q_i = \text{constant} \qquad (1.36)$$

For a trivial example: we are working in cartesian coordinates so that $q = (x, y, z)$. The function $\varphi = (x + \epsilon, y, z)$ describes a transformation in which the entire system is displaced a distance ϵ in the x direction. We say that the Lagrangian is invariant under such a transformation if

$$L(\varphi(t, \epsilon), \dot{\varphi}(t, \epsilon), t) = L(q(t), \dot{q}(t), t) \qquad (1.37)$$

If Equation (1.37) holds then

$$\frac{\partial L}{\partial \epsilon} = \sum_i \left[\frac{\partial L}{\partial \dot{\varphi}_i} \frac{\partial \dot{\varphi}}{\partial \epsilon} + \frac{\partial L}{\partial \varphi_i} \frac{\partial \varphi_i}{\partial \epsilon} \right] = 0 \qquad (1.38)$$

The Lagrange equations (1.23) allow us to replace

$$\frac{\partial L}{\partial \varphi_i} = \frac{\partial}{\partial t} \frac{\partial L}{\partial \dot{\varphi}_i} \qquad (1.39)$$

$$\frac{\partial L}{\partial \epsilon} = \sum_i \left[\frac{\partial L}{\partial \dot{\varphi}_i} \frac{\partial \dot{\varphi}}{\partial \epsilon} + \left(\frac{\partial}{\partial t} \frac{\partial L}{\partial \dot{\varphi}_i} \right) \frac{\partial \varphi_i}{\partial \epsilon} \right] \quad (1.40)$$

$$= \sum_i \frac{\partial}{\partial t} \left(\frac{\partial L}{\partial \dot{\varphi}_i} \frac{\partial \varphi_i}{\partial \epsilon} \right) = 0$$

In the limit $\epsilon \to 0$

$$\varphi_i \to q_i \qquad \dot{\varphi}_i \to \dot{q}_i \qquad \frac{\partial}{\partial t} \to \frac{d}{dt} \qquad \frac{\partial L}{\partial \dot{\varphi}_i} = p_i$$

$$\sum_i p_i \delta q_i = \text{constant} \quad (1.41)$$

This is the final statement of Noether's theorem. For every transformation of the sort described above that leaves the Lagrangian invariant, there is a conserved quantity given by (1.41).

Let's look at an example where the answer is known already, circular motion in a plane. The Lagrangian is given by (1.25).

$$L = \frac{1}{2} m(\dot{r}^2 + r^2 \dot{\phi}^2) - V(r)$$

The transformation is $\phi \to \phi + \epsilon$, *i.e.* $\varphi_1 = r$, $\varphi_2 = \phi + \epsilon$, $p_1 = m\dot{r}^2$, $p_2 = mr^2\dot{\phi}$.

$$\delta q_1 = \delta r = \frac{\partial r}{\partial \epsilon} = 0$$

$$\delta q_2 = \delta \phi = \frac{\partial}{\partial \epsilon}(\phi + \epsilon) = 1$$

According to (1.41) the conserved quantity is p_2, which of course is angular momentum.

Despite its fundamental significance, Noether's theorem is not much help in practical calculations. Granted it gives a procedure for finding the conserved quantity *after the corresponding symmetry transformation has been found,* but how is one to find the transformation? The physicist must rely on his traditional tools: inspiration, hard work, and the Ouija Board. The fact remains that there are simple systems, *e.g.* the Henon-Heiles problem to be discussed later, that have fascinated physicists for decades and for which the existence of these transformations is still controversial.

I will have much more to say about this question. As you will see, there is a more or less "cookbook" procedure for finding the right set of variables and some fundamental results about the sorts of problems for which these procedures are possible.

1.6 Velocity-dependent forces and potentials

In deriving equation (1.23) we had to assume that there was no velocity dependence in the potentials. There are two important problems for which this is not true, friction and magnetic forces, but in the case of friction we can derive a close analog to (1.16), and in the case of electromagnetic forces we can redefine the Lagrangian in such a way that it satisfies (1.23) exactly as it is written. Friction is the easier case. Let's deal with it first.

Assume that frictional force is proportional to velocity.[4]

$$\boldsymbol{F} = -k\boldsymbol{v}$$

Define Rayleigh's dissipation function

$$D \equiv \frac{1}{2}k\boldsymbol{v} \cdot \boldsymbol{v}$$

Evidently

$$F_x = -\frac{\partial D}{\partial v_x} \qquad \boldsymbol{F} = -\boldsymbol{\nabla}_v D$$

So far, these are ordinary forces. To complete the derivation we need the generalized force. Using (1.11) and (1.7) we have

$$\Im_j = \boldsymbol{F} \cdot \frac{\partial \boldsymbol{r}}{\partial q_j} = -\boldsymbol{\nabla}_v D \cdot \frac{\partial \boldsymbol{r}}{\partial q_j} = -\boldsymbol{\nabla}_v D \cdot \frac{\partial \dot{\boldsymbol{r}}}{\partial \dot{q}_j} = -\frac{\partial D}{\partial \dot{q}_j}$$

So the equation equivalent to (1.16) is

$$\frac{d}{dt}\left(\frac{\partial T}{\partial \dot{q}_j}\right) - \frac{\partial T}{\partial q_j} + \frac{\partial D}{\partial \dot{q}_j} = 0 \qquad (1.42)$$

The case of an object falling through a viscus fluid under the influence of gravity makes an easy example. Since gravity is an ordinary, *i.e.* non velocity-dependent force we can replace T in (1.42) with L from (1.22)

$$L = \frac{1}{2}mv^2 - gmx \qquad D = \frac{1}{2}kv^2$$

The constant g is the acceleration of gravity. The resulting differential equation

$$m\dot{v} + gmx + kv = 0$$

is linear and easy to solve.

$$v = \left(v_0 - \frac{gm}{k}\right)e^{-kt/m} + \frac{gm}{k}$$

---

[4]For example, according to Stokes' law, for a sphere of radius R moving through a fluid with viscosity η, $k = 6\pi\eta R$.

Evidently there is a terminal velocity $v_\infty = gm/k$.

Now let's look at electromagnetic forces. We will work with Cartesian coordinates, so there is no distinction between ordinary force and generalized force.

$$\mathfrak{F} = \boldsymbol{F} = q[\boldsymbol{E} + \boldsymbol{v} \times \boldsymbol{B}] \tag{1.43}$$

$$\boldsymbol{E} = -\boldsymbol{\nabla}\phi - \frac{\partial \boldsymbol{A}}{\partial t} \qquad \boldsymbol{B} = \boldsymbol{\nabla} \times \boldsymbol{A}$$

Theorem 1.1. *Define a generalized electromagnetic potential*

$$U \equiv q\phi - q\boldsymbol{A} \cdot \boldsymbol{v} \tag{1.44}$$

Then a new sort of Lagrangian defined by

$$L \equiv T - U \tag{1.45}$$

satisfies the Lagrange equation (1.23).

The proof of this theorem requires two simple lemmas.

Lemma 1.1.

$$\frac{d}{dt}A_x = \frac{\partial}{\partial x}(\boldsymbol{A} \cdot \boldsymbol{v}) - (\boldsymbol{v} \times \boldsymbol{B})_x$$

Proof.

$$\boldsymbol{v} \times \boldsymbol{B} = \boldsymbol{v} \times (\boldsymbol{\nabla} \times \boldsymbol{A}) = \boldsymbol{\nabla}(\boldsymbol{A} \cdot \boldsymbol{v}) - (\boldsymbol{v} \cdot \boldsymbol{\nabla})\boldsymbol{A}$$

$$\frac{\partial}{\partial x}(\boldsymbol{A} \cdot \boldsymbol{v}) = (\boldsymbol{v} \times \boldsymbol{B})_x + (\boldsymbol{v} \cdot \boldsymbol{\nabla})A_x$$

And the same for the y and z components evidentally. Note that the $\boldsymbol{\nabla}$ doesn't differentiate $\boldsymbol{v}$ in the first expression.

$$\frac{d}{dt}A_x = \frac{\partial A_x}{\partial x}\frac{dx}{dt} + \frac{\partial A_x}{\partial y}\frac{dy}{dt} + \frac{\partial A_x}{\partial z}\frac{dz}{dt} = (\boldsymbol{v} \cdot \boldsymbol{\nabla})A_x$$

$$= \frac{\partial}{\partial x}(\boldsymbol{A} \cdot \boldsymbol{v}) - (\boldsymbol{v} \times \boldsymbol{B})A_x$$

$\square$

Lemma 1.2.

$$\mathfrak{F}_k = -\frac{\partial U}{\partial q_k} + \frac{d}{dt}\left(\frac{\partial U}{\partial \dot{q}_k}\right) \tag{1.46}$$

Proof.

$$\frac{\partial U}{\partial \dot{x}} = -q A_x$$

$$\frac{d}{dt}\left(\frac{\partial U}{\partial \dot{x}}\right) = -q\frac{\partial}{\partial x}(\boldsymbol{A} \cdot \boldsymbol{v}) + q(\boldsymbol{v} \times \boldsymbol{B})_x \qquad (1.47)$$

$$-\frac{\partial U}{\partial x} = q E_x + q\frac{\partial}{\partial x}(\boldsymbol{A} \cdot \boldsymbol{v}) \qquad (1.48)$$

Combining the (1.47) and (1.48) gives the desired result.

$$\frac{d}{dt}\left(\frac{\partial U}{\partial \dot{x}}\right) - \frac{\partial U}{\partial x} = q E_x + q(\boldsymbol{v} \times \boldsymbol{B})_x = \Im_x \qquad (1.49)$$

$\square$

Proof of Theorem 1.1. Combine (1.45), (1.49), and (1.16). The $\Im_k$.s cancel leaving the desired result, Equation (1.23). $\square$

We have fulfilled the terms of our contract. The potential U defined by (1.44) satisfies (1.16) with $\Im_x$ defined by (1.43). We are then guaranteed that L defined by (1.45) will satisfy Lagrange's equation.

1.7 The Hamiltonian formulation

I will explain the Hamiltonian assuming that there is only one degree of freedom. It's easy to generalize once the basic ideas are clear. Lagrangians are functions of q and $\dot{q}$. We define a new function of q and p given by (1.34).

$$H(p, q) = p\,\dot{q} - L(q, \dot{q}) \qquad (1.50)$$

The new function is called the Hamiltonian, and the transformation $L \to H$ is called a Legendre transformation. The equation is much more subtle than it looks. In fact, it's worth several pages of explanation.

It's clear from elementary mechanics that q, $\dot{q}$, and p can't all be independent variables, since p is usually a simple function of $\dot{q}$. You might say that there are two ways of formulating Newton's second law: a $(q, \dot{q})$ formulation, $F = m\ddot{q}$, and a (q, p) formulation, $F = \dot{p}$. The connection between q and it's canonically conjugate momentum is usually more complicated than this, but there is still a $(q, \dot{q})$ formulation, the Lagrangian, and a (q, p) formulation, the Hamiltonian. The Legendre transformation is a procedure for transforming the one formulation into the other. The key point is that it is invertible.[5] To see what this means, let's first assume that q, $\dot{q}$ and p are all independent.

$$H(q, \dot{q}, p) = p\,\dot{q} - L(q, \dot{q}) \tag{1.51}$$

$$dH = \left(p - \frac{\partial L}{\partial \dot{q}}\right) d\dot{q} + \dot{q}\,dp - \frac{\partial L}{\partial q}\,dq \tag{1.52}$$

What is the condition that H *not* depend on $\dot{q}$?

$$\boxed{p(q, \dot{q}) = \frac{\partial L(q, \dot{q})}{\partial \dot{q}}} \tag{1.53}$$

This is the definition of p anyhow, so we're on the right track.

$$dH = \dot{q}\,dp - \frac{\partial L}{\partial q}\,dq$$

$$dH = \frac{\partial H}{\partial p}\,dp + \frac{\partial H}{\partial q}\,dq$$

Adding and subtracting these two equations gives

$$\boxed{\dot{q}(q, p) = \frac{\partial H}{\partial p}} \tag{1.54}$$

$$\boxed{-\frac{\partial L}{\partial q} = \frac{\partial H}{\partial q}} \tag{1.55}$$

Combining (1.23), (1.53), and (1.55) gives the fourth major result.

$$\boxed{\dot{p}(q, p) = -\frac{\partial H}{\partial q}} \tag{1.56}$$

Now here's what I mean that Legendre transformations are invertible: First follow the steps from $L \to H$. We start with $L = L(q, \dot{q})$. Equation (1.53) gives $p = p(q, \dot{q})$. Invert this to find $\dot{q} = \dot{q}(q, p)$. The Hamiltonian is now

$$H(q, p) = \dot{q}(q, p)p - L[q, \dot{q}(q, p)]. \tag{1.57}$$

Now suppose that we start from $H = H(p, q)$. Use (1.54) to find $\dot{q} = \dot{q}(q, p)$. Invert to find $p = p(q, \dot{q})$. Finally

$$L(q, \dot{q}) = \dot{q}p(q, \dot{q}) - H[q, p(q, \dot{q})] \tag{1.58}$$

---

[5]The following argument is taken from Hand and Finch, [Hand and Finch (1998)].

In both cases we were able to complete the transformation *without knowing ahead of time the functional relationship among* q, $\dot{q}$, *and* p. To summarize: Equations (1.51), (1.53), and (1.55) enable us to transform between the $(q, \dot{q})$ (Lagrangian) prescription and the (q, p) (Hamiltonian) prescription; while (1.54) and (1.56) are Hamilton's equations of motion.

There is one more foundational equation that illustrates the role of conservation of energy.

$$H = \dot{q}\frac{\partial L}{\partial \dot{q}} - L$$

$$\frac{dH}{dt} = \ddot{q}\frac{\partial L}{\partial \dot{q}} + \dot{q}\frac{d}{dt}\left(\frac{\partial L}{\partial \dot{q}}\right) - \frac{dL}{dt}$$

but

$$\frac{dL}{dt} = \frac{\partial L}{\partial q}\dot{q} + \frac{\partial L}{\partial \dot{q}}\ddot{q} + \frac{\partial L}{\partial t}$$

Combining these last two equations gives the final result.

$$\boxed{\frac{dH}{dt} = -\frac{\partial L}{\partial t}} \tag{1.59}$$

This is the bottom line; *unless the Lagrangian depends explicitly on time, H is constant.*

1.7.1 *The spherical pendulum*

A mass m hangs from a string of length R. The string makes an angle θ with the vertical and can rotate about the vertical with an angle ϕ.

$$T = \frac{1}{2}mR^2(\dot{\theta}^2 + \sin^2\theta\,\dot{\phi}^2) \tag{1.60}$$

$$V = mgR(1 - \cos\theta) \tag{1.61}$$

The mgR constant doesn't appear in the equations of motion, so we can forget about it. The Lagrangian is $L = T - V$ as usual.

$$p_\theta = \frac{\partial L}{\partial \dot{\theta}} = mR^2\dot{\theta} \tag{1.62}$$

$$p_\phi = \frac{\partial L}{\partial \dot{\phi}} = mR^2\sin^2\theta\,\dot{\phi} \equiv l_\phi \tag{1.63}$$

The angle ϕ is cyclic, so $p_\phi = l_\phi$ is constant. At this point we are still in the $(q, \dot{q})$ prescription. Invert (1.62) and (1.63) to obtain $\dot{\theta}$ and $\dot{\phi}$ as functions of p_θ and l_ϕ.

$$\dot{\theta} = p_\theta / mR^2 \tag{1.64}$$

$$\dot{\phi} = l_\phi / mR^2 \sin^2 \theta \tag{1.65}$$

$$H = \frac{p_\theta^2}{2mR^2} + \frac{l_\phi^2}{2mR^2 \sin^2 \theta} - mgR \cos \theta \tag{1.66}$$

The equations of motion follow from this.

$$\dot{\theta} = \frac{\partial H}{\partial p_\theta} = \frac{p_\theta}{mR^2} \tag{1.67}$$

$$\dot{p}_\theta = -\frac{\partial H}{\partial \theta} = \frac{l_\phi^2 \cos \theta}{mR^2 \sin^3 \theta} - mgR \sin \theta \tag{1.68}$$

$$\dot{\phi} = \frac{\partial H}{\partial p_\phi} = \frac{l_\phi}{mR^2 \sin^2 \theta} \tag{1.69}$$

$$\dot{p}_\phi = 0 \tag{1.70}$$

Suppose we were to try to find an analytic solution to this system of equations. First note that there are two constants of motion, the angular momentum l_ϕ, and the total energy $E = H$.

(1) Invert (1.66) to obtain $p_\theta = p_\theta(\theta, E, l_\phi)$.
(2) Substitute p_θ into (1.64) and integrate

$$\int \frac{mR^2}{p_\theta} d\theta = t \equiv N(\theta)$$

The integral is hopeless anyhow, so we label its output $N(\theta)$, (short for an exceedingly nasty function).
(3) Invert the nasty function to find θ as a function of t.
(4) Take the sine of this even nastier function and substitute it into (1.69) to find $\dot{\phi}$.
(5) Integrate and invert to find ϕ as a function of t.

This makes sense in principle, but is wildly impossible in practice. Now suppose we could change the problem so that both θ and ϕ were cyclic so that the two constants of motion were p_θ and p_ϕ (rather than E and p_ϕ). Then

$$\dot{\theta} = \frac{\partial H}{\partial p_\theta} = \omega_\theta \qquad \theta = \omega_\theta t + \theta_0$$

$$\dot\phi = \frac{\partial H}{\partial p_\phi} = \omega_\phi \qquad \phi = \omega_\phi t + \phi_0$$

Here ω_θ and ω_ϕ are two constant "frequencies" that we could easily extract from the Hamiltonian. This apparently small change makes the problem trivial! In both cases there are two constants of motion: it makes all the difference which two constants. This is the basis of the idea we will be pursuing in the next chapter.

1.8 Summary

Solving mechanics problems almost always involves exploiting some symmetry; so problems with cylinders are solved in cylindrical coordinates, and when the iconic block slides down the incline plane, it does so in cartesian coordinate systems. Most problems, unfortunately, contain symmetries that don't match any of the familiar coordinate systems, and so much of advanced mechanics is concerned with finding and exploiting such symmetries. The first step in doing this is to reformulate the equations of motion in generalized coordinates and in terms of energies which are scalars, rather with forces which are vectors that must be described with specific coordinate systems. To this end we developed the Lagrangian formulation in terms of coordinates and velocities, q and $\dot q$, Lagrangians, $L(q, \dot q)$, and the Lagrange equation. This is largely a stepping stone to the more powerful Hamiltonian formulation, but it did allow us to give a precise definition of symmetry. A coordinate q that does not appear explicitly in the Lagrangian is said to be cyclic or ignorable. Every such variable indicates the presence of a sort of symmetry, and according to Noether's theorem, a conserved quantity that will simplify the problem substantially. Unfortunately, these symmetries may be quite obscure and only apparent when just the right set of coordinates are used. How to find such symmetries and the corresponding coordinates will occupy us in the next chapter, but the Lagrangian formulation is not well-suited to this task. It is better to formulate the equations of motion in terms of coordinates and momenta, q and p, the Hamiltonian, $H(q, p)$, and Hamilton's equations of motion. The reason is that there is a well-developed body of theory regarding the transformation of Hamiltonians from one coordinate system to another. This is the topic of the next chapter.

1.9   Sources and references

Goldstein, H., Poole, C., and Safko, J. (2002). *Classical Mechanics* (Addison Wesley).

Hand, L. N. and Finch, J. D. (1998). *Analytic Mechanics* (Cambridge University Press).

Taylor, J. R. (2005). *Classical Mechanics* (University Science Books).

1.10   Problems

(1) A bead with mass m slides along a frictionless wire in a uniform gravitational field so that $F_z = g$. The curvature of the wire is completely known.

  (a) Write down the equations of motion in cartesian coordinates using $F = ma$. You may use any functional form you wish to describe the curvature of the wire. Don't forget the forces of constraint.

  (b) Now use the generalized coordinate q which is distance measured along the wire. Assume that the curvature of the wire can be expressed as $z = z(q)$ where $z(q)$ is some known function. Write down the equation of motion.

  (c) Do you get the point?

(2) The equation for the motion of a pendulum is

$$\ddot{\theta} = -\omega^2 \sin\theta$$

  (a) Show that $I_1 \equiv \frac{1}{2}(\dot{\theta}^2 - \omega^2 \cos\theta)$ is a constant. This is sometimes called the *first integral*.

  (b) Use this to derive

$$t = \int_{\theta_0}^{\theta} \frac{d\theta'}{\sqrt{2(I_1 + \omega^2 \cos\theta')}} \equiv [F(\theta')]_{\theta_0}^{\theta}$$

  (c) Define $F(\theta_0) \equiv I_2$. This is a constant called the *second integral* which evidentally depends on the initial conditions. Show that the final solution is

$$\theta(t) = F^{-1}(t + I_2)$$

  In this equation F^{-1} is the inverse function of F. (Not $1/F$)

  (d) Now approximate, $\sin\theta \approx \theta$. Show that you get the expected result for the simple pendulum.

(3) This procedure is sometimes called *the method of quadratures*. (This is potentially misleading since the method of quadratures usually refers to some sort of approximation scheme.) It's a kind of "cook book" procedure that requires no thought to complete, but – the first integral requires elliptic functions, and I have no idea how to invert it to get the second integral. The technique is mostly of theoretical interest; a problem that can be solved in principle using the method of quadratures can be solved in principle. So what are the limits of the method? To explore this question consider the more general problem

$$\ddot{x} = F(x)$$

Where $F(x)$ is a more-or-less arbitrary function except that it must have no singularities, no explicit time dependence, and no dependence on $\dot{x}$. Show that this problem can always be solved using the method of quadratures. What does the final solution look like? Can this method be generalized to two or more independent variables?

(4) Here is the Lagrangian for the driven harmonic oscillator. $F(t)$ is some arbitrary function of time.

$$L = \frac{1}{2}(\dot{q}^2 - \omega^2 q^2) + F(t)q$$

Can you find the solution with the method of quadratures? (Hint: No) What goes wrong? What is the significance of what goes wrong?

(5) A bead of mass m slides without friction in a uniform gravitational field on a vertical circular hoop of radius R as shown in Figure 1.1. The hoop is constrained to rotate at a fixed angular velocity Ω about its vertical diameter. Let θ be the position of the bead on the hoop measured from the lowest point.

    (a) Write down the Lagrangian $L(\theta, \dot{\theta})$.

    (b) Find how the equilibrium values of θ depend on Ω. Which are stable and which are unstable? Note that equilibrium points are those where the acceleration is zero. What happens when the system is displaced slightly from equilibrium? If the forces are such as to return the system to equilibrium it is said to be a stable point. Otherwise it is an unstable point. Take the pendulum for example. There are two equilibrium points: straight down (stable) and straight up (unstable).

(6) Two particles of different masses m_1 and m_2 are connected by a massless spring of spring constant k and equilibrium length d. The system

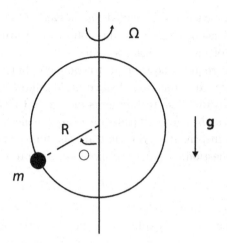

Fig. 1.1   The bead sliding on a rotating hoop.

rests on a frictionless table and may translate, oscillate, and rotate. Find Lagrange's equations of motion. Are there any ignorable coordinates? What are the conjugate momenta? Find the Hamiltonian and Hamilton's equations of motion.

(7) When we were discussing electromagnetic forces, I introduced the notion of a generalized potential defined by (1.44). This is the quantity U in equation (1.46) repeated here for convenience.

$$\Im_j = -\frac{\partial U}{\partial q_j} + \frac{d}{dt}\left(\frac{\partial U}{\partial \dot{q}_j}\right)$$

($\Im_j$ is the generalized force.) With this in mind work the following problem:

A particle moves in a plane under the influence of a central force, whose magnitude is

$$F = \frac{K}{r^2}\left(1 - \frac{\dot{r}^2 - 2\ddot{r}r}{c^2}\right)$$

(This formula has some historical significance. It was suggested by Weber as a correction to Coulomb's law due accelerated motion.) Find the generalized potential that will result in such a force, and for that the Lagrangian for the motion in a plane.

(8) Two mass points of mass m_1 and m_2 are connected by a string passing through a hole in a smooth table as shown in Figure 1.2 The mass

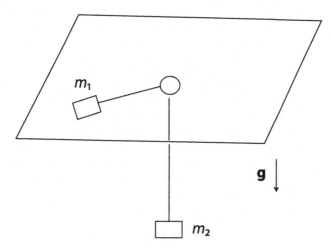

Fig. 1.2   The weight sliding on a table.

m_1 rests on the table surface and m_2 hangs suspended. Assuming m_2 moves only in a vertical line, what are the generalized coordinates of the system? Write Lagrange's equations of motion for the system and if possible discuss the physical significance any of them might have. Reduce the problem to a second-order differential equation and obtain a first integral of the equation. What is it's physical significance?

Chapter 2

Canonical transformations

We saw at the end of the last chapter that a problem in which all the generalized coordinates are cyclic is trivial to solve. We also saw that there is a great flexibility allowed in the choice of coordinates for any particular problem. It turns out that there is an important class of problems for which it is possible to choose the coordinates so that they are in fact all cyclic. The choice is usually far from obvious, but there is a formal procedure for finding the "magic" variables. One formulates the problem in terms of the natural p's and q's and then transforms to a new set of variables, usually called Q_k and P_k, that have the right properties.

2.1 Contact transformations

The most general transformation is called a *contact transformation*.

$$Q_k = Q_k(q, p, t) \qquad P_k = P_k(q, p, t) \qquad (2.1)$$

In this formula and what follows, the symbols p and q when used as arguments stand for the complete set, $q_1, q_2, q_3, \cdots$, *etc.* There is a certain privileged class of transformations called *canonical transformations* that preserve the structure of Hamilton's equation of motion for all dynamical systems. This means that there is a new Hamiltonian function called $K(Q, P)$ for which the new equations of motion are

$$\dot{Q}_k = \frac{\partial K}{\partial P_k} \qquad \dot{P}_k = -\frac{\partial K}{\partial Q_k} \qquad (2.2)$$

In a footnote Goldstein *et al.* [Goldstein *et al.* (2002)] suggest that K be called the Kamiltonian. The idea has caught on with several authors, and I will use it without further apology. The trick is to find it.

Canonical transformations require a generating function conventionally called F. It is a function of one of the original variables, q and p, and one of the new variables, Q and P. Hence there are four choices, $F_1(q, Q, t)$, $F_2(q, P, t)$, $F_3(p, Q, t)$, and $F_4(p, P, t)$. The four are related by Legendre transformations. We will be primarily concerned with F_2, but for pedagogic reasons it's best to start with F_1.

Theorem 1. *Let F_1 be any function of q_k and Q_k and possibly p_k and P_k, as well as time. Then the new Lagrangian defined by*

$$\bar{L} = L - \frac{dF_1}{dt} \tag{2.3}$$

is equivalent to L in the sense that it yields the same equations of motion.

Proof.

$$\dot{F}_1 = \sum_k \frac{\partial F_1}{\partial q_k} \dot{q}_k + \sum_k \frac{\partial F_1}{\partial Q_k} \dot{Q}_k + \frac{\partial F_1}{\partial t} \tag{2.4}$$

$$\frac{d}{dt}\left(\frac{\partial \dot{F}_1}{\partial \dot{q}_k}\right) = \frac{d}{dt}\left(\frac{\partial F_1}{\partial q_k}\right) = \frac{\partial \dot{F}_1}{\partial q_k}$$

$$\frac{d}{dt}\left(\frac{\partial \dot{F}_1}{\partial \dot{Q}_k}\right) = \frac{d}{dt}\left(\frac{\partial F_1}{\partial Q_k}\right) = \frac{\partial \dot{F}_1}{\partial Q_k}$$

These last two can be rewritten

$$\frac{d}{dt}\left(\frac{\partial \dot{F}_1}{\partial \dot{q}_k}\right) - \frac{\partial \dot{F}_1}{\partial q_k} = 0$$

$$\frac{d}{dt}\left(\frac{\partial \dot{F}_1}{\partial \dot{Q}_k}\right) - \frac{\partial \dot{F}_1}{\partial Q_k} = 0$$

So $\dot{F}_1$ satisfies Lagrange's equation whether we regard it as a function of q_k or Q_k. Obviously, if L satisfies Lagrange's equation, then so does $L - \dot{F}_1$. □

K is obtained by a Legendre transformation just as H was.

$$K(Q, P) = \sum_k P_k \dot{Q}_k - \bar{L}(Q, \dot{Q}, t) \tag{2.5}$$

This has the same form as (1.50), so the derivation of the equations of motion (1.53) through (1.56) are unchanged as well.

$$P_k = \frac{\partial \bar{L}}{\partial \dot{Q}_k} \qquad \dot{Q}_k = \frac{\partial K}{\partial P_k} \qquad \dot{P}_k = -\frac{\partial K}{\partial Q_k} \tag{2.6}$$

These simple results provide the framework for canonical transformations. In order to use them we will need to know two more things: how to find F, and given F, how to find the transformation $(q, p) \to (Q, P)$. We deal with the second of these questions now and postpone the first to later sections.

Starting with $F_1(q, Q)$ (2.3) becomes

$$\bar{L}(Q, \dot{Q}, t) = L(q, \dot{q}, t) - \frac{d}{dt} F_1(q, Q, t) \tag{2.7}$$

Since

$$\frac{\partial \bar{L}}{\partial \dot{q}_k} = \frac{\partial L}{\partial \dot{Q}_k} = 0,$$

we get with the help of (2.6)

$$\frac{\partial \bar{L}}{\partial \dot{q}_k} = \frac{\partial L}{\partial \dot{q}_k} - \frac{\partial \dot{F}_1}{\partial \dot{q}_k} = \dot{p}_k - \frac{\partial F_1}{\partial q_k} = 0.$$

$$\frac{\partial \bar{L}}{\partial \dot{Q}_k} = P_k = \frac{\partial L}{\partial \dot{Q}_k} - \frac{\partial \dot{F}_1}{\partial \dot{Q}_k} = -\frac{\partial F_1}{\partial Q_k}$$

This yields the two transformation equations

$$\boxed{P_k = -\frac{\partial F_1}{\partial Q_k}} \tag{2.8}$$

$$\boxed{p_k = \frac{\partial F_1}{\partial q_k}} \tag{2.9}$$

A straightforward set of substitutions gives our final formula for the Kamiltonian.

$$K = \sum_k \left[-\frac{\partial F_1}{\partial Q_k} \dot{Q}_k - L + \frac{\partial F_1}{\partial q_k} \dot{q}_k + \frac{\partial F_1}{Q_k} \dot{Q}_k \right] + \frac{\partial F_1}{\partial t}$$

$$= -L + \sum_k p_k \dot{q}_k + \frac{\partial F_1}{\partial t}$$

To be more explicit

$$\boxed{K(Q, P) = H(q(Q, P), p(Q, P), t) + \frac{\partial}{\partial t} F_1(q(Q, P), Q, t)} \tag{2.10}$$

Since F can be any function (so long as it has the right variables) we have a wide range of possibilities. Our first goal will be to find a transformation such that Q is cyclic and consequently P is a constant of the motion.

Summary:

(1) Here is the typical problem; we are given the Hamiltonian $H = H(q, p)$ for some conservative system. $H = E$ is constant, but the q's and p's change with time in a complicated way. Our goal is to find the functions $q = q(t)$ and $p = p(t)$ using the technique of canonical transformations.

(2) We need to know the generating function $F = F_1(q, Q)$. This is the hard part, and I'm postponing it as long as possible.

(3) Substitute F into (2.8) and (2.9). This gives a set of coupled algebraic equations for q, Q, p, and P. They must be combined in such a way as to give $q_k = q_k(Q, P)$ and $p_k = p_k(Q, P)$.

(4) Use (2.10) to find K. If we had the right generating function to start with, Q will be cyclic, *i.e.* $K = K(P)$. The equations of motion are obtained from (2.6). $\dot{P}_k = 0$ and $\dot{Q}_k = \omega_k$. The ω's are a set of constants as are the P's. $Q_k(t) = \omega_k t + \alpha_k$. The α's are constants obtained from the initial conditions.

(5) Finally $q_k(t) = q_k(Q(t), P)$ and $p_k(t) = p_k(Q(t), P)$.

2.1.1   The harmonic oscillator: cracking a peanut with a sledgehammer

$$H = \frac{p^2}{2m} + \frac{kq^2}{2} = \frac{1}{2m}(p^2 + m^2\omega^2 q^2) \qquad (2.11)$$

It's useful to try a new technique on an old problem. As it turns out, the generating function is

$$F_1 = \frac{m\omega q^2}{2}\cot Q \qquad (2.12)$$

The transformation is found from (2.8) and (2.9).

$$p = \frac{\partial F}{\partial q} = m\omega q \cot Q$$

$$P = -\frac{\partial F}{\partial Q} = \frac{m\omega q^2}{2\sin^2 Q}$$

Solve for p and q in terms of P and Q and then substitute into (2.10) to find K.

$$q = \sqrt{\frac{2P}{m\omega}}\sin Q \qquad p = \sqrt{2Pm\omega}\cos Q$$

$$K = \omega P \qquad P = E/\omega$$

We have achieved our goal. Q is cyclic, and the equations of motion are trivial.

$$\dot{Q} = \frac{\partial K}{\partial P} = \omega \qquad Q = \omega t + Q_0 \qquad (2.13)$$

$$q = \sqrt{\frac{2E}{m\omega^2}}\,\sin(\omega t + Q_0) \qquad p = \sqrt{2mE}\,\cos(\omega t + Q_0) \qquad (2.14)$$

2.2 The second generating function

There's an old recipe for tiger stew that begins, "First catch the tiger." In our quest for the tiger, we now turn our attention to the second generating function, $F_2 = F_2(q, P, t)$. F_2 is obtained from F_1 by means of a Legendre transformation.[1]

$$F_2(q, P) = F_1(q, Q) + \sum_k Q_k P_k \qquad (2.15)$$

We are looking for transformation equations analogous to (2.8) and (2.9). Since $L = \bar{L} + \dot{F}_1$,

$$\sum_k p_k \dot{q}_k - H = \sum_k P_k \dot{Q}_k - K + \frac{d}{dt}(F_2 - \sum Q_k P_k)$$

$$= -\sum Q_k \dot{P}_k - K + \dot{F}_2$$

Substitute

$$\dot{F}_2 = \sum_k \left[\frac{\partial F_2}{\partial q_k} \dot{q}_k + \frac{\partial F_2}{\partial P_k} \dot{P}_k \right] + \frac{\partial F_2}{\partial t}$$

$$-H = -K + \sum_k \left[\left(\frac{\partial F_2}{\partial q_k} - p_k \right) \dot{q}_k + \left(\frac{\partial F_2}{\partial P_k} - Q_k \right) \dot{P}_k \right] + \frac{\partial F_2}{\partial t}$$

We are working on the assumption that $\dot{q}$ and $\dot{P}$ are not independent variables. We enforce this by requiring that

$$\boxed{\frac{\partial}{\partial q_k} F_2(q, P) = p_k} \qquad (2.16)$$

$$\boxed{\frac{\partial}{\partial P_k} F_2(q, P) = Q_k} \qquad (2.17)$$

$$\boxed{K(Q, P) = H(q(Q, P), p(Q, P)) + \frac{\partial}{\partial t} F_2(q(Q, P), P)} \qquad (2.18)$$

---

[1]When in doubt, do a Legendre transformation.

2.3   Hamilton's principal function

The F_1 style generating functions were used to transform to a new set of variables $(q, p) \to (Q, P)$ such that all the Q's were cyclic. As a consequence, the P's were constants of the motion, and the Q's were linear functions of time. The generating function itself was hard to find, however. The F_2 generating function goes one step further; it can transform to a set of variables in which *both* the Q's and P's are constant and simple functions of the initial values of the phase space variables. In essence, our transformation is

$$(q(t), p(t)) \leftrightarrow (q_0, p_0)$$

This is a time dependent transformation, of course. The fact that we can find such transformations shows that the time evolution of a system is itself a canonical transformation.

We look for an F_2 so that K in (2.18) is identically zero! Then from (2.6), $\dot{Q}_k = 0$ and $\dot{P}_k = 0$. The appropriate generating function will be a solution to

$$H(q, p, t) + \frac{\partial F_2}{\partial t} = 0 \tag{2.19}$$

We eliminate p_k using (2.16)

$$H\left(q_1, \ldots, q_n; \frac{\partial F_2}{\partial q_1}, \ldots, \frac{\partial F_2}{\partial q_n}; t\right) + \frac{\partial F_2}{\partial t} = 0 \tag{2.20}$$

The solution to this equation is usually called S, *Hamilton's principal function*. The equation itself is the *Hamilton-Jacobi* equation.[2] There are two serious issues here: does it have a solution, and if it does can we find it? We will take a less serious approach: if we can find a solution, then it most surely exists. Furthermore, the only way I know to solve partial differential equations is by separation of variables. If we succeed the solution will have the form[3]

$$S = \sum_k W_k(q_k) - \alpha t \tag{2.21}$$

Not surprisingly, partial differential equations that have solutions of the form (2.21) are said to be *separable*.[4] Most of the familiar textbook problems in classical mechanics and atomic physics can be separated in this

---

[2]See Goldstein *et al.* [2002], Chapter 10.

[3]Why not simply call it F_2 rather than S? I don't know; this is just standard notation.

[4]Or to be meticulous, *completely separable*.

form. The question of separability does depend on the system of generalized coordinates used. For example, the Kepler problem is separable in spherical coordinates but not in cartesian coordinates. It would be nice to know whether a particular Hamiltonian *could* be separated with *some* system of coordinates, but no completely general criterion is known.[5] As a rule of thumb, Hamiltonians with explicit time dependence are *not* separable.

If our Hamiltonian is separable, then when (2.21) is substituted into (2.20), the result will look like

$$f_1 \left(q_1, \frac{dW_1}{dq_1} \right) + f_2 \left(q_2, \frac{dW_2}{dq_2} \right) + \cdots = \alpha \qquad (2.22)$$

Each function f_k is a function only of q_k and dW_k/dq_k. Since all the q's are independent, each function must be separately constant. This gives us a system of n independent, first-order, ordinary differential equations for the W_k's.

$$f_k \left(q_k, \frac{dW_k}{dq_k} \right) = \alpha_k \qquad (2.23)$$

The W's so obtained are then substituted into (2.21). The resulting function for S is

$$F_2 \equiv S = S(q_1, \ldots, q_n; \alpha_1, \ldots, \alpha_n; \alpha, t)$$

The final constant α is redundant for two reasons: first, $\sum \alpha_k = \alpha$, and second, the transformation equations (2.16) and (2.17) involve derivatives with respect to q_k and P_k. When S is so differentiated, the $-\alpha t$ piece will disappear. In order to make this apparent, we will write S as follows:

$$F_2 \equiv S = S(q_1, \ldots, q_n; \alpha_1, \ldots, \alpha_n; t) \qquad (2.24)$$

Since the F_2 generating functions have the form $F_2(q, P)$, we are entitled to think of the α's as "momenta," *i.e.* α_k in (2.24) corresponds to P_k in (2.15). In a way this makes sense. Our goal was to transform the time-dependent q's and p's into a new set of constant Q's and P's, and the α's are most certainly constant. On the other hand, they are *not* the initial momenta p_0 that evolve into $p(t)$. The relationship between α and p_0 will be determined later.

If we have dome our job correctly, the Q's given by (2.17) are also constant. They are traditionally called β, so

$$Q_k = \beta_k = \frac{\partial S(q, \alpha, t)}{\partial \alpha_k} \qquad (2.25)$$

Again, β's are constant, but they are not equal to q_0.

We can turn this into a cookbook algorithm.

---

[5]The is a very technical result, the so-called Staeckel conditions, which gives necessary and sufficient conditions for separability in orthogonal coordinate systems.

(1) Substitute (2.21) into (2.20) and separate variables.
(2) Integrate the resulting first-order ODE's. The result will be n independent functions $W_k = W_k(q, \alpha)$. Put the W_k's back into (2.21) to construct $S = S(q, \alpha, t)$.
(3) Find the constant β coordinates using

$$\beta_k = \frac{\partial S}{\partial \alpha_k} \tag{2.26}$$

(4) Invert these equations to find $q_k = q_k(\beta, \alpha, t)$.
(5) Find the momenta with

$$p_k = \frac{\partial S}{\partial q_k} \tag{2.27}$$

2.3.1　The harmonic oscillator: again

The harmonic oscillator provides an easy example of this procedure.

$$H = \frac{1}{2m}(p^2 + m^2\omega^2 q^2)$$

$$\frac{1}{2m}\left[\left(\frac{\partial S}{\partial q}\right)^2 + m^2\omega^2 q^2\right] + \frac{\partial S}{\partial t} = 0$$

$$\frac{1}{2m}\left[\left(\frac{dW}{dq}\right)^2 + m^2\omega^2 q^2\right] = \alpha$$

Since there is only one q, the entire quantity on the left of the equal sign is a constant.

$$W(q, \alpha) = \sqrt{2m\alpha} \int dq \sqrt{1 - \frac{m\omega^2 q^2}{2\alpha}}$$

The new transformed constant "momentum" $P = \alpha$.

$$\beta = \frac{\partial S(q, \alpha, t)}{\partial \alpha} = \frac{\partial W(q, \alpha)}{\partial \alpha} - t$$

$$= \frac{1}{\omega}\sin^{-1}\left[q\sqrt{\frac{m\omega^2}{2\alpha}}\right] - t \tag{2.28}$$

Invert this equation to find q as a function of t and β.

$$q = \sqrt{\frac{2\alpha}{m\omega^2}} \, \sin(\omega t + \beta\omega)$$

Evidently, β has something to do with initial conditions: $\omega\beta = \phi_0$, the initial phase angle.

$$p = \frac{\partial S}{\partial q} = \sqrt{2m\alpha - m^2\omega^2 q^2}$$

$$= \sqrt{2m\alpha} \, \cos(\omega t + \phi_0)$$

The maximum value of p is $\sqrt{2mE}$, so that makes sense too.

2.4   Hamilton's characteristic function

There is another way to use the F_2 generating function to turn a difficult problem into an easy one. In the previous section we chose $F_2 = S = W - \alpha t$, so that $K = 0$. It is also possible to to take $F_2 = W(q)$ so that

$$K = H\left(q_k, \frac{\partial W}{\partial q_k}\right) = E \equiv \alpha_1 \qquad (2.29)$$

The W obtained in this way is called *Hamilton's characteristic function.*

$$W = \sum_k W_k(q_k, \alpha_1, \dots, \alpha_n)$$

$$= W(q_1, \dots, q_n, E, \alpha_2, \dots, \alpha_n) = W(q_1, \dots, q_n, \alpha_1, \dots, \alpha_n) \qquad (2.30)$$

It generates a contact transformation with properties quite different from that generated by S. The equations of motion are

$$\dot{P}_k = -\frac{\partial K}{\partial Q_k} = 0 \qquad (2.31)$$

$$\dot{Q}_k = \frac{\partial K}{\partial P_k} = \frac{\partial K}{\partial \alpha_k} = \delta_{k1} \qquad (2.32)$$

The new feature is that $\dot{Q}_1 = 1$ so $Q_1 = t - t_0$. In general

$$Q_k = \frac{\partial W}{\partial \alpha_k} = \beta_k \qquad (2.33)$$

but now $\beta_1 = t - t_0$.

$$p_k = \frac{\partial W}{\partial q_k} \qquad (2.34)$$

as before.

The algorithm now works like this:

(1) Substitute (2.30) into (2.29) and separate variables.
(2) Integrate the resulting first-order ODE's. The result will be n independent functions $W_k = W_k(q, \alpha)$. Put the W_k's back into (2.30) to construct $S = S(q, \alpha, t)$.
(3) Find the constant β coordinates using

$$\beta_k = \frac{\partial S}{\partial \alpha_k} \qquad (2.35)$$

Remember that $\beta_1 = t - t_0$.
(4) Invert these equations to find $q_k = q_k(\beta, \alpha, t)$
(5) Find the momenta with

$$p_k = \frac{\partial S}{\partial q_k} \qquad (2.36)$$

2.4.1   Examples

Problems with one degree of freedom are virtually identical whether they are formulated in terms of the characteristic function or the principal function. Take for example, the harmonic oscillator from the previous section. Equation (2.28) becomes

$$\beta = \frac{\partial W(q,\alpha)}{\partial \alpha} = \frac{1}{\omega}\sin^{-1}\left[q\sqrt{\frac{m\omega^2}{2\alpha}}\right] = t - t_0$$

$$q = \sqrt{\frac{2\alpha}{m\omega^2}}\,\sin[\omega(t-t_0)] \qquad (2.37)$$

The following problem raises some new issues. Consider a particle in a stable orbit in a central potential. The motion will lie in a plane so we can do the problem in two dimensions.

$$H = \frac{1}{2m}\left(p_r^2 + \frac{p_\psi^2}{r^2}\right) + V(r) \qquad (2.38)$$

$p_\psi = mr^2\dot{\psi}$ is the angular momentum. It is conserved since ψ is cyclic.

$$\frac{1}{2m}\left[\left(\frac{\partial W}{\partial r}\right)^2 + \frac{1}{r^2}\left(\frac{\partial W}{\partial \psi}\right)^2\right] + V(r) = \alpha_1 \qquad (2.39)$$

$$\left[r^2\left(\frac{dW_r}{dr}\right)^2 + 2mr^2V(r) - 2m\alpha_1 r^2\right] + \left(\frac{dW_\psi}{d\psi}\right)^2 = 0 \qquad (2.40)$$

At this point we notice $\partial W/\partial \psi = p_\psi$, which we know is constant. Why not call it something like α_ψ? Then $W_\psi = \alpha_\psi \psi$. This is worth stating as a general principle: if q is cyclic, $W_q = \alpha_q q$, where α_q is one of the n constant α's appearing in (2.30).

$$W = \int dr\sqrt{2m(\alpha_1 - V) - \alpha_\psi^2/r^2} + \alpha_\psi \psi \qquad (2.41)$$

We can find r as a function of time by inverting the equation for β_1, just as we did in (2.37), but more to the point

$$\beta_\psi = \frac{\partial W}{\partial \alpha_\psi} = -\int \frac{\alpha_\psi dr}{r\sqrt{2m(\alpha_1 - V) - \alpha_\psi^2/r^2}} + \psi \qquad (2.42)$$

Make the usual substitution, $u = 1/r$.

$$\psi - \beta_\psi = -\int \frac{du}{\sqrt{2m(\alpha_1 - V(r))/\alpha_\psi^2 - u^2}} \qquad (2.43)$$

This is a new kind of equation of motion, which gives $\psi = \psi(r)$ or $r = r(\psi)$ (assuming we can do the integral), *i.e.* there is no explicit time dependence. Such equations are called *orbit equations*. Often it will be more useful to have the equations in this form, when we are concerned with the geometric properties of the trajectories.

2.5   Action-angle variables

We are pursuing a route to chaos that begins with periodic or quasi-periodic systems . A particularly elegant approach to these systems makes use of a variant of Hamilton's characteristic function. In this technique, the integration constants α_k appearing directly in the solution of the Hamilton-Jacobi equation are not themselves chosen to be the new momenta. Instead, we define a set of constants I_k, which form a set of n independent functions of the α's known as *action variables*. The coordinates conjugate to the I's are angles that increase linearly with time. You are familiar with a system that behaves just like this, the harmonic oscillator!

$$q = \sqrt{\frac{2E}{k}} \sin \psi \qquad p = \sqrt{2mE} \cos \psi$$

Where $\psi = \omega t + \psi_0$. In the language of action-angle variables $I = E/\omega$, so

$$q = \sqrt{\frac{2I}{m\omega}} \sin \psi \qquad p = \sqrt{2mI\omega} \cos \psi$$

I is the "momentum" conjugate to the "coordinate" ψ.

Action-angle variables are only appropriate to bounded motion, but within this limitation they are surprisingly versatile. In fact, all systems that are integrable (a concept I will explain in the next chapter) can be reduced to a set of uncoupled harmonic oscillators. To see what "periodic motion" implies, have a look at the simple pendulum.

$$H = \frac{p_\theta^2}{2ml^2} - mgl \cos \theta = E = \alpha \qquad (2.44)$$

$$p_\theta = \pm \sqrt{2ml^2(E + mgl \cos \theta)} \qquad (2.45)$$

There are two kinds of motion possible. If E is small, the pendulum will reverse at the points where $p_\theta = 0$. The motion is called *libration*, *i.e.* bounded and periodic. If E is large enough, however, the pendulum will swing around a complete circle. Such motion is called *rotation* (obviously). There is a critical value of $E = mgl$ for which, in principle, the pendulum could stand straight up motionless at $\theta = \pi$. An orbit in p_θ - θ phase space corresponding to this energy forms the dividing line between the two kinds of motion. Such a trajectory is called a *separatrix*.

For either type of periodic motion, we can introduce a new variable I designed to replace α as the new constant momentum.

$$I(\alpha) = \frac{1}{2\pi} \oint p(q, \alpha)\, dq \qquad (2.46)$$

This is a definite integral taken over a complete period of libration or rotation.[6]

Theorem 2.1. *The angle ψ conjugate to I is cyclic.*

Proof. Since $I = I(\alpha)$ and $H = \alpha$, it follows that H is a function of I only. $H = H(I)$.

$$\boxed{\dot{I} = -\frac{\partial H}{\partial \psi} = 0} \qquad \boxed{\dot{\psi} = \frac{\partial H}{\partial I} = \omega(I)} \qquad (2.47)$$

$\square$

Theorem 2.2. $\Delta\psi = 2\pi$ *corresponds to one complete cycle of the periodic motion.*

Proof. We are using an F_2 type generating function, which is a function of the old coordinate and new momentum. Hamilton's characteristic function can be written as

$$W = W(q, I). \qquad (2.48)$$

The transformation equations are

$$\boxed{\psi = \frac{\partial W}{\partial I}} \qquad \boxed{p = \frac{\partial W}{\partial q}} \qquad (2.49)$$

Note that

$$\frac{\partial \psi}{\partial q} = \frac{\partial}{\partial I}\left(\frac{\partial W}{\partial q}\right)$$

so

$$\oint d\psi = \oint \frac{\partial \psi}{\partial q}\, dq = \frac{\partial}{\partial I} \oint \frac{\partial W}{\partial q}\, dq = \frac{\partial}{\partial I} \oint p\, dq = \frac{\partial}{\partial I}(2\pi I) = 2\pi.$$

$\square$

---

[6]Textbooks are about equally divided on whether to call action I or J and whether or not to include the factor $1/2\pi$.

2.5.1 *The harmonic oscillator: yet again*

$$H = \frac{1}{2m}(p^2 + m^2\omega^2 q^2)$$

$$p = \pm\sqrt{2mE - m^2\omega^2 q^2}$$

$$I = \frac{1}{2\pi} \oint \sqrt{2mE - m^2\omega^2 q^2}\, dq$$

The integral is tricky in this form because p changes sign at the turning points. We won't have to worry about this if we make the substitution

$$q = \sqrt{\frac{2E}{m\omega^2}} \sin\psi \qquad (2.50)$$

This substitution not only makes the integral easy and takes care of the sign change, it also makes clear the meaning of an integral over a complete cycle, *i.e.* ψ goes from 0 to 2π.

$$I = \frac{E}{\pi\omega} \oint \cos^2\psi\, d\psi = E/\omega$$

From this point of view the introduction of ψ at (2.50) seems nothing more that a mathematical trick. We would have stumbled on it eventually, however, as the following argument shows. Simply solve the Hamilton-Jacobi equation.

$$\frac{1}{2m}\left[\left(\frac{dW}{dq}\right)^2 + m^2\omega^2 q^2\right] = E$$

$$W = \int \sqrt{2mI\omega - m^2\omega^2 q^2}\, dq$$

$$\frac{\partial W}{\partial I} = m\omega \int \frac{dq}{\sqrt{2mI\omega - m^2\omega^2 q^2}}$$

$$= \sin^{-1}\left(q\sqrt{\frac{m\omega^2}{2I}}\right) - \psi_0 = \psi$$

$$q = \sqrt{\frac{2E}{m\omega^2}} \sin(\psi - \psi_0)$$

In the last equation ψ_0 appears as an integration constant. Evidently, ψ is the angle variable conjugate to I.

In summary, to use action-angle variables for problems with one degree of freedom:

(1) Find p as a function of $E = \alpha$ and q.
(2) Calculate $I(E)$ using (2.46).
(3) Solve the Hamilton-Jacobi equation to find $W = W(q, I)$.
(4) Find $\psi = \psi(q, I)$ using (2.49).
(5) Invert this equation to get $q = q(\psi, I)$.
(6) Use (2.47) to get $\omega(I)$.
(7) Calculate $p = p(I, q)$ from (2.49).

One attractive feature of this scheme is that you can find the frequency without using the characteristic function and without finding the equations of motion. The phase space plot is particularly important. Use polar coordinates (what else) for (I, ψ). Every trajectory, whatever the system, is a circle!

Our derivation was based on the following assumptions: (1) The system had one degree of freedom. (2) Energy was conserved and the Hamiltonian had no explicit time dependence. (3) The motion was periodic.[7] Every such system is at heart, a harmonic oscillator. Phase space trajectories are circles. The frequency can be found with a few deft moves. From a philosophical point of view, (and we will be getting deeper and deeper into philosophy as these lectures proceed) problems in this category are "as good as solved," nothing more needs to be said about them. The same is definitely not true true with more than one degree of freedom. I will take a paragraph to generalize before going on to some more abstract developments.

We must assume that the system is separable, so

$$W(q_1, \ldots, q_n, \alpha_1, \ldots, \alpha_n) = \sum_k W_k(q_k, \alpha_1, \ldots, \alpha_n) \qquad (2.51)$$

$$p_k = \frac{\partial}{\partial q_k} W_k(q_k, \alpha_1, \ldots, \alpha_n) \qquad (2.52)$$

$$I_k = \frac{1}{2\pi} \oint p_k(q_k, \alpha_1, \ldots, \alpha_n) \qquad (2.53)$$

Next find all the q's as function of the I's and substitute into W.

$$W = W(q_1, \ldots, q_n; I_1, \ldots, I_n)$$

Finally

$$\psi_k = \frac{\partial W}{\partial I_k} \qquad \dot{I}_k = 0 \qquad \dot{\psi}_k = \frac{\partial H}{\partial I_k} = \omega_k \qquad (2.54)$$

[7]With two or more degrees of freedom it is sometimes useful to distinguish between periodic systems in which all the angle variables return to their starting values simultaneously, *i.e.* the entire system is periodic, and conditionally periodic systems in which each angle independently returns to its starting value.

2.6 Summary

Mechanics problems tend to fall into one of two categories; they are either trivial or impossible, with those in the latter category far outnumbering those in the former. Perhaps for every seemingly impossible problem there is some system of coordinates in terms of which the problem becomes trivial. Equations (2.3) and (2.5) give us hope. *Every* function F can be used to create a new system of coordinates and a new Hamiltonian (conventionally called K) which keep Hamilton's equations intact. This function is called the generating function; the problem is to find the right one. So solving these seeming impossible problems comes down to finding the solution to one of two closely related, first order, nonlinear, partial differential equations, (2.20) or (2.29). The only general purpose technique for solving such equations is to separate variables as in (2.21), but equations for which such solutions can be found are the exception rather than the rule. Maybe these equations can't be solved, not because we are not smart enough, but because they have no solutions. This challenges our accustomed way of thinking. Our equations describe a physical system. If it is given some initial conditions, *something* will happen. How can it be that this something cannot be described by mathematics? In the special case of periodic motion with one degree of freedom, none of these difficulties appear. There is a simple cookbook procedure for introducing action-angle coordinates in terms of which the system looks like a simple harmonic oscillator.

When going from one degree of freedom to two, profound difficulties appear at several levels. First there is the issue of separability. If we cannot solve (2.20) or (2.29) we are stopped "dead in our tracks." Even if we succeed there is the difficult question of how the various degrees of freedom are interrelated. This is the topic of the next chapter. Finally, one might try to solve such problems with perturbation theory and get at least a first-order approximation. But here one faces the possibility of divergent series related to the problem of small devisors. We will discuss this in Chapter 4.

At this point the natural question is why bother? Why not just let the computer integrate the equations of motion numerically. The answer is as simple as it is profound, some equations have no solutions in the sense that every attempted perturbation series will diverge, and if the equations have no solution, the computer will not be able to find it.

2.7   Sources and references

Goldstein, H., Poole, C., and Safko, J. (2002). *Classical Mechanics* (Addison Wesley).

Hand, L. N., and Finch, J. D. (1998). *Analytic Mechanics*, (Cambridge University Press).

José, J. V. and Saletan, E. J. (1998). *Classical Dynamics: A Contemporary Approach* (Cambridge University Press).

Matzner, R. A. and Shepley, L. C. (1991). *Classical Mechanics*, (Prentice-Hall).

Taylor, J. R. (2005). *Classical Mechanics* (University Science Books).

2.8   Problems

(1) A particle in a gravitational field has the Hamiltonian

$$H = \frac{p^2}{2m} + mgz = E$$

Use Hamilton'a characteristic function to find $z(t)$.

(2) A particle slides on a pair of frictionless planes, which therefore make a kind of oscillator as shown in the Figure 2.1.

    (a) Solve the action-angle problem for this situation and display the frequencies.

    (b) Initially the displacement is x_0. What are the energy and frequency of the oscillation?

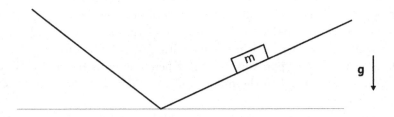

Fig. 2.1   The sliding block oscillator.

(3) Consider an unperturbed one-dimensional Hamiltonian system with

$$H_0 = \frac{p^2}{2m} + V(q)$$

with

$$V(q) = k|q|$$

The motion in such a potential is periodic; note that the period is a function of energy.

(a) Find the equations of motion using the Hamiltonian formalism, *i.e.* find $q(t)$ and $p(t)$.

(b) Solve the problem in Action-angle variables.

(4) Let R, θ, and ϕ be the coordinates of a three-dimensional spherical pendulum. R is the length of the pendulum, θ is its deviation from the vertical axis, and ϕ is its azimuthal coordinate. The Hamiltonian is

$$H = \frac{p_\theta^2}{2mR^2} + \frac{p_\phi^2}{2mR^2 \sin^2 \theta} + mgR(1 - \cos \theta)$$

$$p_\theta = mR^2 \frac{d\theta}{dt} \qquad p_\phi = mR^2 \sin^2 \theta \frac{d\phi}{dt}$$

(a) What are the constants of motion?

(b) Solve the problem with action-angle variables. Find integral expressions for the two angle variables.

(5) A particle of mass m moves in a plane in a square well potential.

$$V(r) = -V_0 \qquad 0 < r < r_0$$

$$= 0 \qquad r > r_0$$

(a) Under what conditions will the motion be bounded and periodic?

(b) Use action-angle variables to find the period of the motion.

(6) In Section 5.9 we discuss the problem of two pendulums coupled by a weak spring. In general the motion is chaotic, but in the limit of small oscillations, an exact solution is possible.

(a) Find the solution using Hamilton's characteristic function.

(b) Express the results using action-angle variables.

(c) Why does the action-angle formalism fail for larger oscillations?

Chapter 3

Abstract transformation theory

So, one-dimensional problems are simple. Given the restrictions listed in the previous section, their phase space trajectories are circles. How does this generalize to problems with two or more degrees of freedom? A brief answer is that, given a number of conditions that we must discuss carefully, the phase space trajectories of a system with n degrees of freedom, move on the surface of an n-dimensional torus imbedded in $2n$-dimensional space. The final answer is a donut! In order to prove this remarkable assertion and understand the conditions that must be satisfied, we must slog through a lot of technical material about transformations in general.

3.1 Notation

Our first job is to devise some compact notation for dealing with higher dimensional spaces. I will show you the notation in one dimension. It will then be easy to generalize. Recall Hamilton's equations of motion.

$$\dot{p} = -\frac{\partial H}{\partial q} \qquad \dot{q} = \frac{\partial H}{\partial p}$$

We will turn this into a vector equation.

$$\boldsymbol{\eta} = \begin{pmatrix} q \\ p \end{pmatrix} \qquad \boldsymbol{J} = \begin{pmatrix} 0 & 1 \\ -1 & 0 \end{pmatrix} \qquad \boldsymbol{\nabla} = \begin{pmatrix} \partial/\partial q \\ \partial/\partial p \end{pmatrix} \qquad (3.1)$$

The equations of motion in vector form are

$$\dot{\boldsymbol{\eta}} = \boldsymbol{J} \cdot \boldsymbol{\nabla} H \qquad (3.2)$$

J is not a vector of course. Equation (3.2) is an example of dyadic notation; it is just shorthand for matrix multiplication, *i.e.*

$$\begin{pmatrix} \dot{q} \\ \dot{p} \end{pmatrix} = \begin{pmatrix} 0 & 1 \\ -1 & 0 \end{pmatrix} \begin{pmatrix} \partial H/\partial q \\ \partial H/\partial p \end{pmatrix}$$

The structure of J is important. Notice that it does two things: it exchanges p and q and it changes one sign. This is called a *symplectic* transformation. I want to explore the connection between canonical transformations and symplectic transformations.

I'll start with the generic canonical transformation, $(q,p) \rightarrow (Q,P)$. How do the velocities transform? Define

$$M = \begin{pmatrix} \partial Q/\partial q & \partial Q/\partial p \\ \partial P/\partial q & \partial P/\partial p \end{pmatrix} \tag{3.3}$$

$$\begin{pmatrix} \dot{Q} \\ \dot{P} \end{pmatrix} = \begin{pmatrix} \partial Q/\partial q & \partial Q/\partial p \\ \partial P/\partial q & \partial P/\partial p \end{pmatrix} \begin{pmatrix} \dot{q} \\ \dot{p} \end{pmatrix} \tag{3.4}$$

Using the notation

$$\zeta = \begin{pmatrix} Q \\ P \end{pmatrix} \tag{3.5}$$

this can be written

$$\dot{\zeta} = M \cdot \dot{\eta} = M \cdot J \cdot \nabla_{(q,p)} H \tag{3.6}$$

The gradient operator differentiates H with respect to q and p. These derivatives transform *e.g.*

$$\frac{\partial H}{\partial q} = \frac{\partial H}{\partial Q}\frac{\partial Q}{\partial q} + \frac{\partial H}{\partial P}\frac{\partial P}{\partial q}$$

consequently

$$\nabla_{(q,p)} = M^T \cdot \nabla_{(Q,P)} H(Q,P) \tag{3.7}$$

The T stands for transpose, of course:

$$\dot{\zeta} = M \cdot J \cdot M^T \cdot \nabla_{(Q,P)} H(Q,P) \tag{3.8}$$

but

$$\dot{\zeta} = J \cdot \nabla_{(Q,P)} H(Q,P) \tag{3.9}$$

Combining (3.8) and (3.9):

$$\boxed{J = M \cdot J \cdot M^T}$$ (3.10)

Those of you who have studied special relativity should find (3.10) congenial. Remember the definition of a Lorentz transformation: any 4×4 matrix Λ that satisfies

$$g = \Lambda \cdot g \cdot \Lambda^T$$ (3.11)

is a Lorentz transformation.[1] The matrix

$$g = \begin{pmatrix} 1 & 0 & 0 & 0 \\ 0 & -1 & 0 & 0 \\ 0 & 0 & -1 & 0 \\ 0 & 0 & 0 & -1 \end{pmatrix}$$ (3.12)

is called the *metric* or *metric tensor*. Forgive me for exaggerating slightly: everything there is to know about special relativity flows out of (3.12). We say that Lorentz transformations "preserve the metric," *i.e.* leave the metric invariant. The geometry of space and time is encapsulated in (3.12). By the same token, canonical transformations preserve the metric J. The geometry of phase space is encapsulated in the definition of J. Since J is symplectic, canonical transformations are *symplectic transformation*, they preserve the symplectic metric.

Equation (3.10) is the starting point for the modern approach to mechanics that uses the tools of Lie group theory. I will only mention in passing some points of contact with group theory. Both [Goldstein *et al.* (2002)] and [Scheck (1994)] have much more on the subject.

3.2 Poisson brackets

Equation (3.10) is really shorthand for four equations, two equation of the form

$$\frac{\partial Q}{\partial q}\frac{\partial P}{\partial p} - \frac{\partial P}{\partial q}\frac{\partial Q}{\partial p} = 1$$ (3.13)

and two trivial equations

$$\frac{\partial Q}{\partial q}\frac{\partial Q}{\partial p} - \frac{\partial Q}{\partial p}\frac{\partial Q}{\partial q} = 0$$

---

[1]It is not a good idea to use matrix notation in relativity because of the ambiguity inherent in covariant and contravariant indices. Normally one would write (3.10) using tensor notation.

$$\frac{\partial P}{\partial q}\frac{\partial P}{\partial p} - \frac{\partial P}{\partial p}\frac{\partial P}{\partial q} = 0$$

This combination of derivatives appears so frequently it is given a special name. The usual notation is

$$\frac{\partial X}{\partial q}\frac{\partial Y}{\partial p} - \frac{\partial X}{\partial q}\frac{\partial Y}{\partial p} \equiv [X,Y]_{q,p} \qquad (3.14)$$

The quantity on the right is called the *Poisson bracket*. Then (3.13) becomes

$$[Q,P]_{q,p} = 1 \qquad (3.15)$$

This together with the trivially true

$$[q,p]_{q,p} = 1 \qquad (3.16)$$

are called the *fundamental Poisson brackets*.

This can be easily generalized to n dimensions.

$$\boldsymbol{\eta} = \begin{pmatrix} q_1 \\ q_2 \\ \vdots \\ q_n \\ p_1 \\ p_2 \\ \vdots \\ p_n \end{pmatrix} \qquad \boldsymbol{J} = \begin{pmatrix} 0 & I_n \\ -I_n & 0 \end{pmatrix} \qquad \boldsymbol{\nabla} = \begin{pmatrix} \partial/\partial q_1 \\ \partial/\partial q_2 \\ \vdots \\ \partial/\partial q_n \\ \partial/\partial p_1 \\ \partial/\partial p_2 \\ \vdots \\ \partial/\partial p_n \end{pmatrix} \qquad (3.17)$$

The symbol I_n is the $n \times n$ unit matrix. The Poisson bracket is now defined as follows.

$$[X,Y]_\eta \equiv \sum_k \left(\frac{\partial X}{\partial q_k}\frac{\partial Y}{\partial p_k} - \frac{\partial X}{\partial p_k}\frac{\partial Y}{\partial q_k} \right) \qquad (3.18)$$

or in matrix notation

$$[X,Y]_\eta = (\boldsymbol{\nabla}_\eta X)^T \cdot \boldsymbol{J} \cdot \boldsymbol{\nabla}_\eta Y \qquad (3.19)$$

If we choose X and Y to be one or the other of the q's and p's we have

$$[q_i, q_k]_\eta = [p_i, p_k]_\eta = 0 \qquad [q_i, p_k]_\eta = \delta_{ik} \qquad (3.20)$$

These are of course the commutation relations for position and momentum operators in quantum mechanics. The resemblance is not accidental. The operator formulation of quantum mechanics grew out of the Poisson bracket formulation of classical mechanics. This development is reviewed in all the standard texts.

We can make the notation more compact by defining a matrix Poisson bracket $[\boldsymbol{X}, \boldsymbol{Y}]$ whose ij element is $[X_i, Y_j]$. In this notation the fundamental Poisson bracket is

$$[\boldsymbol{\eta}, \boldsymbol{\eta}]_\eta = \boldsymbol{J} \tag{3.21}$$

It is an important fact that the Poisson bracket is invariant under canonical transformations, *i.e.*

$$[X, Y]_\eta = [X, Y]_\zeta \tag{3.22}$$

The proof is straightforward.

$$\boldsymbol{\nabla}_\eta Y = \boldsymbol{M}^T \cdot \boldsymbol{\nabla}_\zeta Y$$

$$(\boldsymbol{\nabla}_\eta X)^T = (\boldsymbol{M}^T \cdot \boldsymbol{\nabla}_\zeta X)^T = (\boldsymbol{\nabla}_\zeta X)^T \cdot \boldsymbol{M}$$

$$[X, Y]_\eta = (\boldsymbol{\nabla}_\zeta X)^T \cdot \boldsymbol{M} \cdot \boldsymbol{J} \cdot \boldsymbol{M}^T \cdot \boldsymbol{\nabla}_\zeta Y$$

$$= (\boldsymbol{\nabla}_\zeta X)^T \cdot \boldsymbol{J} \cdot \boldsymbol{\nabla}_\zeta Y = [X, Y]_\zeta$$

The last step makes use of (3.10). The invariance of the Poisson brackets is a non-trivial consequence of the symplectic nature of canonical transformations. From now on will will not bother with the subscripts on the Poisson brackets.

Here is another similarity with quantum mechanics. Let f be any function of canonical variables.

$$\dot{f} = \sum_k \left(\frac{\partial f}{\partial q_k} \dot{q}_k + \frac{\partial f}{\partial p_k} \dot{p} \right) + \frac{\partial f}{\partial t}$$

$$= \sum_k \left(\frac{\partial f}{\partial q_k} \frac{\partial H}{\partial p_k} - \frac{\partial f}{\partial p_k} \frac{\partial H}{\partial q_k} \right) + \frac{\partial f}{\partial t}$$

$$\boxed{\frac{df}{dt} = [f, H] + \frac{\partial f}{\partial t}} \tag{3.23}$$

This looks like Heisenberg's equation of motion. For our purposes it means that if f doesn't depend on time explicitly and if $[f, H] = 0$, then f is a constant of the motion. We can use (3.23) to test if our favorite function is in fact constant, and perhaps we can also use it to construct new constants as the following argument shows.

Let f, g, and h be arbitrary functions of canonical variables. The following *Jacobi identity* is just a matter of algebra.

$$[f, [g, h]] + [g, [h, f]] + [h, [f, g]] = 0 \qquad (3.24)$$

Now suppose $h = H$, the Hamiltonian, and f and g are constants of the motion. Then

$$[H, [f, g]] = 0$$

Consequence: *if f and g are constants of the motion, then so is $[f, g]$.* This should make us uneasy. Take any two constants. Well, maybe they commute, but if not, then we have three constants. Commute the new constant with f and g and get two more constants, *etc.* How many constants are we entitled to – anyway? This is a deep question, which has something to do with the notion of *involution.* I'll get to that later.

3.3   Liouville's theorem

The content of Liouville's theorem is easy to understand if we compare it with the similar theorem in fluid mechanics. Imagine a volume V of moving fluid enclosed by a surface S. As the fluid moves S will change shape and perhaps the volume will change also. The question is, how fast will the volume change? To calculate this, take a small patch on S with area dA. The fluid at this point is moving with velocity $\boldsymbol{v}$. In time δt, the area dA will sweep out a small volume, $\delta V = \boldsymbol{n} \cdot \boldsymbol{v}\, \delta t\, dA$, where $\boldsymbol{n}$ is a normal vector to the surface at this point. Integrating over the surface gives the rate of change.

$$\frac{dV}{dt} = \int_S \boldsymbol{n} \cdot \boldsymbol{v}\, dA \qquad (3.25)$$

If the fluid is incompressible $dV/dt = 0$ and the divergence theorem gives our final result

$$\int_V \boldsymbol{\nabla} \cdot \boldsymbol{v}\, dV = 0 \qquad (3.26)$$

These ideas can be applied to phase space with some important modifications. First, the coordinates of phase space are q and p, so the $\boldsymbol{v}$ is $(\dot{q}, \dot{p})$. For a system with n degrees of freedom, the space is $2n$-dimensional with coordinates (q_i, p_i), $i = 1 \ldots n$. Real fluid is composed of molecules, but there are so many of them that we can consider the fluid to be a continuous medium. In the same way, our phase space "fluid" consists of a vast number of points, each one represents a possible set of (q, p) coordinates of some dynamical system. Since these points are purely conceptual you can populate V with as many as you like. As time elapses these points sweep out trajectories. I am going to prove that so long as Hamilton's equations of motion are satisfied, this swarm of particles moves like an incompressible gas. I will prove this first for one degree of freedom, and consider generalizations later.

The first point to notice is that these points follow trajectories that can't cross. If they did there would be one moment in time where one set of (q, p) coordinates gave birth to two separate trajectories. This is impossible; the laws of mechanics are deterministic; one set of initial conditions uniquely determines all subsequent motion. By the same token, a trajectory inside S can't escape, and one outside can't get in. We can now reproduce the argument leading up to (3.26), this time in two dimensions. Take the two-dimensional curve S in Figure 3.1 It surrounds an area A, and it encloses a swarm of trajectories at time t. The curve S' encloses the same trajectories at time δt later. I will show that the areas enclosed by these two curves are equal. The area between S and S' can be broken up into small parallelograms with sides of length dl and $v\delta t$ where v is the speed with which the surface is moving at that point. One such parallelogram is shown shaded in Figure 3.1. Choose dl and δt so that the velocities are essentially constant over the area of the parallelogram. The area of this patch is then $n \cdot v\, dl\, \delta t$ where n is the unit vector normal to the curve. Integrating this around S gives the total change in area.

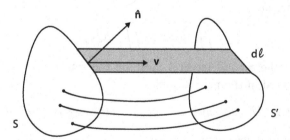

Fig. 3.1   The surface S encloses a family of trajectories at time t. At some time δt later the trajectories (indicated by the curved lines) have evolved so that they are enclosed by the surface S'.

Theorem 3.1.

$$\delta A = \delta t \oint dl\, \boldsymbol{n} \cdot \boldsymbol{v} = 0 \tag{3.27}$$

Proof. From the divergence theorem

$$\oint dl\, \boldsymbol{n} \cdot \boldsymbol{v} = \iint_A \boldsymbol{\nabla} \cdot \boldsymbol{v}\, dA \tag{3.28}$$

We can think of S as a locus of points in phase space taken at time t, each point moving according to (3.2). The trajectories it encloses satisfy the same equations. Using the notation from (3.1)

$$\boldsymbol{v} = \dot{\boldsymbol{\eta}} = \boldsymbol{J} \cdot \boldsymbol{\nabla} H \tag{3.29}$$

It's easy to see that this is zero

$$\boldsymbol{\nabla} \cdot \dot{\boldsymbol{\eta}} = \frac{\partial \dot{q}}{\partial q} + \frac{\partial \dot{p}}{\partial p} = \frac{\partial}{\partial q}\left(\frac{\partial H}{\partial p}\right) + \frac{\partial}{\partial p}\left(-\frac{\partial H}{\partial q}\right) = 0 \tag{3.30}$$

since the order of differentiation doesn't matter. Another way of stating this is that the density of points around any one trajectory is constant. $\square$

These two statements are the Hamiltonian version of (3.26). The motion of trajectories obeying this constraint is often called *Hamiltonian flow*. Because the minus sign in (3.30) is crucial, it is also called *symplectic flow* in honor of $\boldsymbol{J}$ in Equation (3.29). It seems not possible to formulate a similar theorem with the Lagrangian variables $(q, \dot{q})$. This is one more advantage of the Hamiltonian formulation of mechanics over the Lagrangian version.

These ideas are directly relevant to canonical transformations. Consider the transformation $(q, p) \to (Q, P)$. The new variables are supposed to describe the same system as the old, therefore they must have the same phase volume, *i.e.*

$$\iint_V dq\, dp = \iint_{V'} dQ\, dP \tag{3.31}$$

the limits of the second integral must be the same as those of the first except expressed in the new variables.

Theorem 3.2. *The Jacobian of a canonical transformation is unity.*

Proof. This is almost obvious since

$$\iint dq\, dp = \iint \frac{(\partial q, \partial p)}{(\partial Q, \partial P)} dQ\, dP = \iint dQ\, dP \tag{3.32}$$

$\square$

Notice that if (Q, P) simply described the same trajectories as (q, p) only at a later time, *i.e.* $Q(t) = q(t+\delta t)$, the truth of (3.32) would be guaranteed by Theorem 3.1. The fact that it is also guaranteed by Theorem 3.2 leads to an important conclusion, *the time evolution of a system is itself a canonical transformation*.

Theorem 3.2 provides a useful test of whether or not a transformation is canonical. For example, the rather odd transformation

$$q \to P \qquad p \to -Q$$

is in fact canonical. It seems that, so far as transformations go, what we call position and what we call momentum are mostly a matter of convenience.

Liouville's theorem also gives an alternate route to derive some of the results from Chapter 2. For example

$$\iint dp\, dq = \iint dP\, dQ \text{ or } \oint p\, dq = \oint P\, dQ \qquad (3.33)$$

The last equality follows from Stokes's theorem. (More about this later.) Now suppose we consider Q and q as the independent variables so that $P = P(Q, q)$ and $p = p(Q, q)$.

$$\oint [p(Q, q)dq - P(Q, q)dQ] = 0 \qquad (3.34)$$

The quantity in square brackets must be some differential, let's call it $dF_1(Q, q)$,

$$\oint dF_1(Q, q) = \oint \left(\frac{\partial F_1}{\partial q}dq + \frac{\partial F_1}{dQ}dQ \right) \qquad (3.35)$$

Comparing (3.34) and (3.35) gives

$$\frac{\partial F_1}{\partial q} = p \qquad \frac{\partial F_1}{\partial Q} = -P \qquad (3.36)$$

as we derived in Section 2.1, Equations (2.8) and (2.9). If we had chosen the independent variables to be P and q this same argument would have yielded (2.16) and (2.17).

What about higher-dimensional spaces? The proof can be generalized to n dimensions.[2] The result, as you might expect, is

$$\int_V d\boldsymbol{q} \cdot d\boldsymbol{p} = \text{ constant} \qquad (3.37)$$

---

[2][Tabor (1989)], Appendix 2.2.

but the interpretation of the integral is far from obvious. To be more explicit we should write

$$\oint \sum_{i=1} p_i dq_i = \sum_{i=1} \int \int_{A_i} dp_i dq_i = \text{constant} \qquad (3.38)$$

What is conserved is the sum of areas of the flux tube projected onto each of the (q_i, p_i) planes at constant time.

In the case of two degrees of freedom it's possible to prove a stronger version of this theorem. This will be extremely important when we study Poincaré sections, so I will explain the theorem in some detail. The proof can be found in [Tabor (1989)].[3] Consider a two degree of freedom system with phase space coordinates (p_x, x, p_y, y) and define an initial curve $\mathcal{C}$ as a set of initial conditions on the (p_x, x) plane at $y = 0$ and at constant energy so that p_y is automatically specified. Under Hamiltonian flow these points will move through phase space sweeping out a tube of trajectories. If the system is bounded and conservative, the tube will eventually recross the $y = 0$ plane marking out a new closed curve $\mathcal{C}'$. There is no reason to believe that all the trajectories will cross the $y = 0$ plane at the same time; nevertheless, *the areas enclosed by $\mathcal{C}$ and $\mathcal{C}'$ are equal,* i.e.

$$\oint_{\mathcal{C}} p_x dx = \oint_{\mathcal{C}'} p_x dx \qquad (3.39)$$

Figure 3.2 is an attempt to portray all four dimensions on a two-dimensional piece of paper! We will make crucial use of this in Chapter 5.[4]

We have seen that the time evolution of a system is canonical and that canonical transformations preserve phase space volume. This simple general statement allows us to prove a remarkable theorem due to Poincaré. The theorem applies to energy-conserving Hamiltonian systems with bounded orbits. Roughly speaking, almost every system, given enough time, will return arbitrarily close to its starting configuration an infinite number of times![5] Here is a heuristic proof. Imagine a starting volume of phase space. As time goes by this constant volume will wonder through phase space sweeping out a "tube" (called a phase tube). But since the phase space is bounded, the tube will eventually have to intersect itself. The points in the intersection region originally came from the starting volume, so that must have been the initial intersection point. This violates the second law of thermodynamics as well as common sense. Take for example a container that is partitioned so that one half is filled with gas and the

---

[3] *ibid.* Appendix 4.1.

[4] This result can be generalized to higher-dimensional spaces. *ibid.* Appendix 4.1.

[5] "In our end is our beginning." T. S. Eliot, *Burnt Norton* from *Four Quartets*.

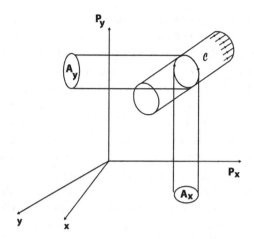

Fig. 3.2 The contour $\mathcal{C}$ encloses an area that is evolving in four dimensions (p_x, x, p_y, y). The integral in (3.38) consists of the sum of A_x, the projection of the area enclosed by $\mathcal{C}$ on the plane $p_y = 0, y = 0$, and A_y, the projection of the area enclosed by $\mathcal{C}$ on the plane $p_x = 0, x = 0$.

other half is evacuated. Remove the partition and the gas rushes into the vacuum until equilibrium is reached. According to the theorem the gas will eventually go back where it came from and will continue to revisit its place of origin throughout eternity. This is the Poincaré recurrence theorem. A more formal proof follows.

Proof. Let U be a neighborhood in phase space and T a time development operator that steps the system forward in time one "click" at a time. The sequence

$$U, \ TU, \ T^2 U, \ldots, T^n U, \ldots$$

contains successive images of U all with the same volume. If they never intersected our phase space would have to have infinite volume, therefore for some $k \geq 0$ and $l \geq 0$ with $k > l$ the union

$$T^k U \bigcap T^l U \neq \emptyset$$

Now step the system backward l times.

$$T^{k-l} U \bigcap U \neq \emptyset$$

So if y is in this union, then $y = T^{k-l} x$ with $x \in U$, i.e. x and $T^{k-l} x$ are both contained in U. $\qquad \square$

Some comments are in order. Remember that x represents an entire system, not just a single molecule for example. The argument shows that some systems return to their starting volume, but do they all? Consider those x's in U but not in $T^{k-l}U$. They will also sweep out a flux tube and we can repeat the previous argument to show that some of these at least will return. We can repeat the argument as often as we like to catch the "strays" but we can not guarantee that none escape. That is the reason for the caveat "almost" in the statement of the theorem. I've said nothing about the size of U. It can be as small as you like, but it must be finite. That is the reason for the caveat "arbitrarily close."

The theorem gives us no way to estimate how long x will take to return except to say that the smaller U the longer it will take. Obviously in our example of the container of gas, the recurrence time would be longer than the age of the universe, so in that sense the theorem is of no consequence. It is theoretically significant however. For one thing it shows that the second law of thermodynamics can be broken. Any "proof" that the second law "must" be true is suspect. Another theoretical point of great interest has to do with the direction of time. It is paradoxical that the microscopic laws of physics are time reversal symmetric[6] and yet in the macroscopic world time flows like a river – never up hill! What is it that defines the arrow of time? One suggestion is that it is the second law of thermodynamics; time moves in the direction in which entropy increases. But if entropy sometimes decreases, what then? If you were part of a system that was in the processes of recurring, would you grow younger? Would your eggs unbreak and your cakes unbake?[7]

3.4   Geometry in n dimensions: the hairy ball

Have another look at equation (3.2). With n degrees of freedom it defines a vector field $\dot{\eta}$ at each point in the $2n$-dimensional phase space. Since energy is conserved in the problems we are considering, the constraint $H = $ constant defines a $2n - 1$ dimensional manifold on which the

---

[6]There is actually a small violation of time reversal symmetry in the world of elementary particles. It appears in the K meson system. Rarely the long-lived kaon metamorphoses in to the short-lived kaon. This seems fantastically obscure, but if there were no violation of time reversal symmetry all the particles and antiparticles would have annihilated in the early universe and there would be nothing left but photons and neutrinos. Philosophers and cosmologists who discuss the nature of time tend to regard this as insignificant. This may be pure chauvinism, but I take the existence of matter very seriously.

[7]An extended discussion of these issues can be found in [Carroll (2010)].

trajectories are confined. The gradient of a function has a well-defined geometric significance; at the point P on the smooth surface $f = \text{constant}$, $\boldsymbol{\nabla} f$ is perpendicular to the flat surface tangent to the surface of constant f at the point P. The vector $\dot{\boldsymbol{\eta}}$ is perpendicular to this, since according to (3.19)

$$(\boldsymbol{\nabla} f)^T \cdot \boldsymbol{J} \cdot \boldsymbol{\nabla} f = [f, f] = 0$$

This is easy to visualize when $n = 1$. The trajectories are closed and periodic. A canonical transformation can transform the (q, p) variables to action-angle variables (I, ψ). In geometric terms, the original trajectories are deformed or mapped into circles; $\dot{\boldsymbol{\eta}}$ is everywhere tangent to the circles.

When $n = 2$ an interesting new problem emerges. Suppose there is an additional constant of the motion so the trajectories are confined to a two-dimensional manifold embedded in four-dimensional space. What is the shape of this manifold? Our experience with the action-angle variables $(I_1, \psi_1, I_2, \psi_2)$ suggests that the manifold can be transformed into a sphere, but this turns out to be impossible. There is a well-known theorem from algebraic topology called the Poincaré-Hopf theorem or in more familiar terms, the hairy ball theorem, which proves that it is impossible to define a smooth vector field on a sphere without having the field equal to zero at some point. The idea is this: try to comb the hair on a hairy ball so that there is no bald spot. It can't be done. So long as you really use a *comb*, *i.e.* so long as the trajectories don't cross, you will always be left with one hair standing straight up! This is not a proof, of course, but it is a vivid way of visualizing the content of the theorem.[8] Another way to look at it is to consider the lines of latitude and longitude on the surface of the earth. Every point of the earth can be associated with a unique latitude and longitude – except for the poles which have every possible longitude. Is it possible to come up with some new coordinate system in which this problem doesn't arise? No.[9] There are other problems with spheres which are symptomatic. There is only one constant of motion, the radius; and every curve can be continuously deformed into any other curve. For example, any longitude can be transformed into a latitude.

---

[8]Topologists have had a field day restating this theorem. "You can't comb a hairy ball flat without creating a cowlick." "You can't comb the hair on a coconut." "Every cow must have at least one cowlick." These colorful descriptions are all taken from Wikipedia. A more prosaic (and useful) statement is that every smooth vector field on a sphere has a singular point.

[9]The hairy ball theorem only applies to spheres with an even number of dimensions. Obviously there is no problem combing a hairy circle.

It is easy to see, however, that none of these problems arise on the surface of a torus. A torus has two radii. Trajectories going around the torus the "long way" can't be deformed into trajectories going around the "short way" without cutting through the donut. Finally if we choose our longitudes so they go around the torus the long way and latitudes so they go around the short way there will be no singular points.

In introducing the $n = 2$ problem I glibly assumed that there were two constants of motion. This assumption needs to be examined more carefully. Suppose the second constant was just a function of the first or that the two constants were functions of a third. It seems intuitively clear that this would not provide enough constraints to solve the problem. Somehow the two constants must be independent. The notion of functional independence can be formulated in several ways. I will present two which are easy to understand when there are just two (putatively) independent constants.

Given two constants of motion called F and G, one of which might be the Hamiltonian. Particle trajectories must lie on the three-dimensional manifold defined by $F = $ constant and simultaneously on the $G = $ constant manifold. If there is to be a unique trajectory it must lie on the intersection of these two manifolds, and this intersection must have dimension two, *i.e.* a line imbedded in four-dimensional space. This is easy to generalize to higher dimensions. Suppose there are n degrees of freedom and n constants of motion, $F_i = $ constant, $i = i \ldots n$. Then the intersection of all these manifolds must have dimension n.

For each of these constants of motion we can construct a vector field using (3.2).

$$\dot{\boldsymbol{\eta}}_F = \boldsymbol{J} \cdot \boldsymbol{\nabla} F$$

$$\dot{\boldsymbol{\eta}}_G = \boldsymbol{J} \cdot \boldsymbol{\nabla} G$$

It should be possible to use these vectors to make an unambiguous local coordinate system. We do this as follows. Take a point P anywhere on the intersection of the constant F and G surfaces and a flat plane tangent to F centered at P. We need two infinitesimal vectors, $\hat{\boldsymbol{\xi}}_F = \epsilon \dot{\boldsymbol{\eta}}_F$ and $\hat{\boldsymbol{\xi}}_G = \epsilon \dot{\boldsymbol{\eta}}_G$, such that every trajectory in the $\hat{\boldsymbol{\xi}}_F$ - $\hat{\boldsymbol{\xi}}_G$ plane has constant F and G. $\hat{\boldsymbol{\xi}}_F$ is guaranteed to lie in the surface of constant F; however, G should remain constant along $\hat{\boldsymbol{\xi}}_F$. This means that

$$0 = (\hat{\boldsymbol{\xi}}_G)^T \cdot \boldsymbol{\nabla} F = \epsilon (\boldsymbol{J} \cdot \boldsymbol{\nabla} F)^T \cdot \boldsymbol{\nabla} G$$

$$= -\epsilon (\boldsymbol{\nabla} F)^T \cdot \boldsymbol{J} \cdot \boldsymbol{\nabla} G = -\epsilon [F, G]$$

The result is surprisingly simple, if F and G are independent then $[F, G] = 0$. We say that two commuting constants like this are in *involution* with one another. This result also relieves our anxiety about extra constants. If F and G are independent, we don't get a "free" constant $K = [F, G]$, because $K = 0$.

A problem with two degrees of freedom and a separable Hamiltonian satisfies the above conditions almost trivially. The two action variables I_1 and I_2 are the two radii of the torus and the angles ψ_1 and ψ_2 specify the position of the particle on the torus. The trajectory remains on the torus and the motion is periodic or conditionally periodic. Such problems seem contrived and uninteresting, but this simple example can be generalized to a remarkable extent by what is called the Liouville integrability theorem. I will simply state and explain this very nontrivial theorem. The complete proof can be found in [Arnold (2010)]. A brief outline of the proof appears in [Jose and Saletan (1998)].

Suppose we are given n functions in involution on a $2n$-dimensional manifold.

$$F_1, \ldots, F_n \qquad [F_i, F_j] = 0 \qquad i, j = 1, 2, \ldots, n$$

Consider the set $\mathbf{M_F}$ of all x such that

$$F_i(x) = f_i, \qquad i = 1, \ldots, n$$

with constant f_i. Assume that the n functions are independent on $\mathbf{M_F}$ in all the senses explained above. Then

- $\mathbf{M_F}$ is a smooth manifold on which the trajectories move according to the Hamiltonian $H = F_1$.
- If the manifold $\mathbf{M_F}$ is compact[10] and connected it is diffeomorphic to a n-dimensional torus.[11]
- The motion is periodic or conditionally periodic and can be described with action-angle variables. Systems of this sort are said to be integrable.

The remarkable thing about this theorem is that it is not necessary to assume that the Hamiltonian is separable. I'll finish with a quote from [Arnold (2010)]. "In fact the theorem of Liouville formulated above covers all the problems in dynamics which have been integrated to the present day."

---

[10]There are numerous technicalities surrounding the notion of compact manifolds. For our purposes it just means that the motion is bounded.

[11]There is a continuous invertible mapping between $\mathbf{M_F}$ and a torus, *i.e.* $\mathbf{M_F}$ can be continuously deformed into the shape of a torus.

3.5 Summary

(1) A system with n degrees of freedom has at most n independent constants of motion. Otherwise we could use the additional constants to eliminate one or more of the free variables.

(2) Poisson brackets can be used to identify independent constants of motion, since they commute with the Hamiltonian and with one another.

(3) If there are n independent constants, then subject to some technical requirements, the system is integrable using action-angle variables.

(4) In the case of integrable systems the trajectories are confined to an n-dimensional torus imbedded in $2n$-dimensional phase space.

(5) If there are fewer than n independent constants there are no general statements we can make about the behavior of the trajectories. We will be very much concerned in the next chapter with systems that are "almost" integrable.

(6) There are no general criteria known for deciding whether or not a system has n independent constants and is therefore integrable; however if the Hamiltonian is separable, the system is integrable.

3.5.1 *Example: uncoupled oscillators*

The Hamiltonian for two uncoupled harmonic oscillators (with $m = 1$) is

$$H = \frac{1}{2}(p_1^2 + p_2^2 + \omega_1^2 q_1^2 + \omega_2 q_2^2) \qquad (3.40)$$

This is an important problem because most linear oscillating systems can be put in this form by a suitable choice of coordinates.[12] There are two constants of motion

$$E_1 = \frac{1}{2}(p_1^2 + \omega_1^2 q_1^2) \qquad E_2 = \frac{1}{2}(p_2^2 + \omega_2^2 q_2^2)$$

In terms of action-angle variables, the constants are I_1 and I_2.

$$H = I_1\omega_1 + I_2\omega_2 = E_1 + E_2 = E$$

Every integrable system can be put in this form, although in general the ω's will be functions of the I's. Here they are just parameters from the Hamiltonian.

---

[12]This comes under the heading of "theory of small oscillations." Most mechanics texts devote a chapter to it.

This is a simple problem, but the phase space is four-dimensional. Let's think about all possible ways we might visualize it. In the (q_1, p_1) or (q_2, p_2) plane the trajectories are ellipses with

$$q_k(\text{max}) = \sqrt{2E_k}/\omega_k \qquad p_k(\text{max}) = \sqrt{2E_k}$$

where $k = 1, 2$. The area enclosed by each ellipse is significant, because

$$\text{area} = \int_s dq \, dp = \oint p \, dq = 2\pi I \tag{3.41}$$

The first integral is a surface integral over the area of the ellipse. The second is a line integral around the ellipse. This identity is a variant of Stokes's theorem. It's useful to rescale the variables so that they both have the same units and the trajectory is a circle. An natural choice would be

$$q_k' = q_k \sqrt{\omega_k} = \sqrt{2I_k} \, \sin\psi_k \qquad p_k' = p_k/\sqrt{\omega_k} = \sqrt{2I_k} \, \cos\psi_k$$

The trajectories are now circles with radius $\sqrt{2I_k}$. The area enclosed is $2\pi I_k$, as required by (3.41).

The motion in the (q_1, q_2) plane is more complicated. It depends on the ratio ω_1/ω_2 called the *winding number*. If this is a rational number, say N_1/N_2 then after N_1 cycles of q_1 and N_2 cycles of q_2, the trajectory will come back to its starting point. This is called a Lissajou figure. If the winding number is irrational, the trajectory will be confined to a limited area but will never return to its starting point. It will eventually "color in" all available space. In the next chapter we will be concerned with systems that are "almost" integrable. For such systems the winding number is all-important. Systems with irrational winding numbers tend to be stable under perturbation. Those with rational winding numbers disintegrate at the slightest push!

The centerpiece of this chapter is the torus. If the winding number is rational the trajectories "wear a path" around the torus. If it's irrational they cover the torus evenly. A useful way of visualizing this was invented by Poincaré. Imagine a flat plane cutting through the torus in such a way that every point on the plane has the angle $\psi_1 = 0$. Place a dot on the plane at the point where each trajectory passes through it with $\dot{\psi}_1 > 0$. If the winding number is a rational fraction, there will be a finite number of points. Each time a trajectory passes through $\psi_1 = 0$ it will pass through one of the dots. If the winding number is irrational the crossings will mark out a continuous circle. The Poincaré section as it is called (some books call it the surface of section) is a useful diagnostic tool. Suppose you have a system of equations that are not integrable (so far as you know) but is amenable to computer calculation. Take various Poincaré sections. If they are circles or at least closed curves then the system is at least approximately integrable and can be described with action-angle variables. As we will see, there are often regions of phase space, "islands" as it were, where motion is simply periodic and other regions that are wildly chaotic.

Pictures of this motion appear in all the standard texts, but what does it mean really to say that the torus is a 2-d surface in a 4-d space? Your breakfast donut after all, is imbedded in 3-d space. If we take a Poincaré section through the torus at the plane $\psi_2 = 0$ and plot q_1 versus p_1, we will get either a circle of dots or a continuous circle with a radius equal to $\sqrt{2I_1}$, or we can take a slice through $\psi_1 = 0$ and get a circle with radius $\sqrt{2I_2}$. Put it this way, any point on the torus has *four* (polar) coordinates, $(\sqrt{2I_1}, \psi_1, \sqrt{2I_2}, \psi_2)$, but in 3-d space, only *three* of them are independent. When the torus is in 4-d space, all four of them are independent. If we really lived in 4-d space, we would label the axes of the donut plot (q_1, p_1, q_2, p_2). This is impossible for us to imagine. The donut is easy; just remember that there is no equation of constraint among the four variables.[13]

3.5.2   *Example: a particle in a box*

Consider a particle in a two-dimensional box with elastic walls.

$$0 \leq x \leq a \qquad 0 \leq y \leq b$$

$$H = \frac{1}{2m}(p_x^2 + p_y^2) = \frac{\pi^2}{2m}\left(\frac{I_1^2}{a^2} + \frac{I_2^2}{b^2}\right)$$

$$I_1 = \frac{1}{2\pi}\oint p_x\, dx = \frac{a}{\pi}|p_x| \qquad I_2 = \frac{b}{\pi}|p_y|$$

$$\omega_1 = \frac{\partial H}{\partial I_1} = \frac{\pi^2}{ma^2}I_1 \qquad \omega_2 = \frac{\pi^2}{mb^2}I_2$$

There are several interesting points about this apparently trivial problem. This looks like a linear system, but in fact it contains an invisible nonlinear potential that reverses the particle's momentum when it hits the wall. One symptom of this is that the frequencies depend on I. This looks odd, but it's just the action-angle way of saying that the particle makes a round trip

---

[13]Of course, I_1 and I_2 are constant for any given set of initial conditions. It is this sense in which the torus is a 2-d surface.

(in the x direction) in a time $T = 2am/p_x$. The loop integral in this context is an integral over one "round trip" of the particle.

$$\oint p_x dx = \int_0^a |p_x|\, dx + \int_a^0 (-|p_x|)\, dx = 2a|p_x|$$

My real point in showing this example is to call your attention to the angle variable. I will work through the calculation for the x variable. This same thing holds for y of course.

$$\frac{1}{2m}\left(\frac{dW_x}{dx}\right)^2 = E_1$$

$$W_x = \int (\pm)\sqrt{2mE_1}\, dx = \int (\pm)\frac{\pi I_1}{a}\, dx$$

$$\psi_1 = \frac{\partial W_x}{\partial I_1} = \pm\frac{\pi}{a}\int dx = \pm\frac{\pi}{a}x + \psi_{10} = \psi_1$$

The term ψ_{10} is an integration constant. There is no reason why it must be the same for both legs of the journey. We are free to choose it as follows:

$$0 \to x \to a: \qquad \psi_1 = \pi x/a$$

$$0 \leftarrow x \leftarrow a: \qquad \psi_1 = 2\pi - \pi x/a$$

While the particle is bouncing violently between the walls, the angle variables are increasing smoothly with time, $\psi_1 = \omega_1 t$ and $\psi_2 = \omega_2 t$. Even this strange problem is equivalent to a donut![14]

3.6 Problems

(1) A system with two degrees of freedom is described by the Hamiltonian

$$H = q_1 p_1 - q_2 p_2 - aq_1^2 + bq_2^2$$

Show that

$$F_1 = \frac{p_1 - aq_1}{q_2} \text{ and } F_2 = q_1 q_2$$

are constants of motion. Are there any other independent algebraic constants of motion? Can any be constructed from the Jacobi identity? (Taken from Goldstein, Poole and Safco, (2002))

---

[14]When correctly viewed, everything is a harmonic oscillator – in this case two harmonic oscillators.

(2) Apply Liouville's theorem to the harmonic oscillator. Show that in a small domain in phase space the system moves in a circle like the hands of a clock. Now try the same thing for the pendulum. How does the shape of the domain change as the motion nears the separatrix? You might want to check your results with the figures in Scheck (1994).

(3) (a) In Section 1.6 I showed that Maxwell's equations could be derived from a Lagrangian. Show that they can also be derived from a Hamiltonian.

   (b) It follows that a beam of charged particles must obey Liouville's theorem. This is an important point in the design of particle accelerators. Consider the following simple example. A colliding beam accelerator produces a pulse or "packet" of particles in the shape of a cylinder of length L in the beam direction and radius R. The packet inevitably spreads out as it is accelerated because the particles have a very small spread in longitudinal momentum $p_z \pm \Delta p_z$. It also has a small spread in transverse momentum between zero and $\Delta p_\perp$. As it approaches the interaction region it is focused electromagnetically to increase its density. Assuming that L and Δp_z don't change, what happens to R and $\Delta p_\perp$?

3.7   Sources and references

Arnold, V. I. (2010). *Mathematical Methods of Classical Mechanics, Second Edition* (Springer).

Carroll, Sean (2010). *From Eternity To Here* (Penguin Group).

Goldstein, H., Poole, C., and Safco, J. (2002). *Classial Mechanics, Third Edition* (Addison Wesley).

José, V. J. and Saletan, E. J. (1998). *Classical Dynamics: A Contemporary Approach* (Cambridge University Press).

Matzner, R. A. and Shepley, L. C. (1991). *Classical Mechanics* (Prentice-Hall).

Scheck, F. (1994). *Mechanics*, (Springer-Verlag).

Tabor, M. (1989). *Chaos and Integrability in Nonlinear Dynamics* (John Wiley & Sons).

Chapter 4

Canonical perturbation theory

So far we have assumed that our systems had exact analytic solutions. One way of stating this is that we can find a canonical transformation to action-angle variables such that the new Hamiltonian is a function of the action variables only, $H = H(I)$. Such problems are the exception rather than the rule. For our purposes they are also uninteresting. All periodic integrable systems are equivalent to a set of uncoupled harmonic oscillators. Once you get over the thrill of this discovery, the oscillators are boring! The existence of chaos depends on the system *not* being equivalent to a set of oscillators. In order to deal with systems that are non-trivial in this sense, we need some way of doing perturbation theory.[1]

4.1 One-dimensional systems

I will present the theory first for systems with one degree of freedom. This will simplify the notation, however the interesting complications only appear in higher dimensions. Here is the basic situation: A bounded conservative system with one degree of freedom is described by a constant Hamiltonian $H(q, p) = E$. We need to obtain the equations of motion in the form $q = q(t)$ and $p = p(t)$, but this is impossible due to the non-linear nature of the problem. We are able to split up the Hamiltonian

$$H = H_0 + \epsilon H_1$$

in such a way that H_0 is amenable to exact solution, and H_1 is in some sense small. We indicate the smallness by multiplying it by ϵ. This is a

---

[1] I will follow the treatment in [Tabor (1989)]. Another good reference is [Matzner and Shepley (1991)]. The subject is also discussed in [Goldstein *et al.* (2002)]. Goldstein *et al.* discussed time-dependent and time-independent perturbation theory. We are doing the time-independent variety.

bookkeeping device; it will be set to one after the approximations have been derived.

The first step is to find the canonical transformation that makes H_0 cyclic, *i.e.* $q = q(\psi, I)$, $p = p(\psi, I)$, and $H_0 = H_0(I)$ where I and $\dot{\psi} = \omega$ are both constant. Unfortunately, this transformation does not render the complete Hamiltonian cyclic, so we write

$$H(\psi, I) = H_0(I) + \epsilon H_1(\psi, I) \tag{4.1}$$

H is still constant, and consequently I now depends on ψ. H_0 is not an explicit function of ψ, but it does depend on ψ implicitly through I.

Despite this inconvenience, I and ψ are still a perfectly good set of canonical variables, so that Hamilton's equations of motion

$$\dot{I} = -\frac{\partial}{\partial \psi} H(\psi, I) \qquad \dot{\psi} = \frac{\partial}{\partial I} H(\psi, I) \tag{4.2}$$

are valid without approximation, even though we are unable to solve them in this form. The so-called time-dependent perturbation proceeds from here by expanding the solutions of (4.2) as power series in ϵ. Our approach is to find a second canonical transformation, *i.e.* $(q, p) \rightarrow (\psi, I) \rightarrow (\varphi, J)$ such that $H(\psi, I) \rightarrow K(J)$. This last step must be done as a series of approximations, of course, otherwise the problem would be exactly solvable.

In order to make the transformation $(\psi, I) \rightarrow (\varphi, J)$ we will use a generating function of the F_2 genus, *i.e.* $F = F(\psi, J)$. For F_2-style functions in general

$$\frac{\partial F(\psi, J)}{\partial \psi} = I \qquad \frac{\partial F(\psi, J)}{\partial J} = \varphi \tag{4.3}$$

We need to expand

$$F = F_0(\psi, J) + \epsilon F_1(\psi, J) + \cdots \tag{4.4}$$

where $F_0 = J\psi$.[2] This is the identity transformation as can be seen as follows:

$$I = \frac{\partial}{\partial \psi} F_0 = J \qquad \varphi = \frac{\partial}{\partial J} F_0 = \psi$$

In terms of (4.4) the transformation equations are

$$I = \frac{\partial F}{\partial \psi} = J + \epsilon \frac{\partial F_1(\psi, J)}{\partial \psi} + \cdots \tag{4.5}$$

---

[2]There is a possibility of confusion here. F_1 in (4.4) is the first-order approximation to F, which is a generating function of the F_2 genus.

$$\varphi = \frac{\partial F}{\partial J} = \psi + \epsilon \frac{\partial F_1(\psi, J)}{\partial J} + \cdots \qquad (4.6)$$

Before going on there are some technical points about ψ and J that need to be discussed. When $\epsilon = 0$, ψ is the exact angle variable for the system. This means that we can find p and q as functions of ψ such that p and q return to their original values when $\Delta\psi = 2\pi$. We can in principle invert this transformation to find ψ as a function of p and q.

$$\psi = \psi(q, p) \qquad (4.7)$$

When p and q run through a complete cycle, ψ advances by 2π. When $\epsilon \neq 0$ the orbit will be different from the unperturbed case, but the functional relationship doesn't change, so when p and q run through a complete cycle, we must still have $\Delta\psi = 2\pi$. Of course, the *exact* angle variable will also advance 2π. In summary

$$\Delta\psi = \Delta\varphi = 2\pi \qquad (4.8)$$

for one complete cycle.

The following integrals are all equal because canonical transformations preserve phase space volume.

$$J = \frac{1}{2\pi} \oint p\, dq = \frac{1}{2\pi} \oint J\, d\varphi = \frac{1}{2\pi} \oint I\, d\psi \qquad (4.9)$$

Now integrate (4.5) around one orbit:

$$\frac{1}{2\pi} \oint I\, d\psi = \frac{1}{2\pi} \oint J\, d\psi + \frac{1}{2\pi} \epsilon \oint \frac{\partial F_1}{\partial \psi}\, d\psi + \cdots$$

that is

$$J = J + \frac{1}{2\pi} \epsilon \oint \frac{\partial F_1}{\partial \psi}\, d\psi + \cdots$$

We have just seen that $\Delta\psi = 2\pi$ around one cycle. Consequently

$$\oint \frac{\partial F_1}{\partial \psi}\, d\psi = 0 \qquad (4.10)$$

implies that the derivative of F_1 is purely oscillatory with a fundamental period of 2π in ψ. (The same is true of the higher order terms as well.)

The Hamiltonian is transformed using (2.18) with the new variables.

$$K(\varphi, J) = H(\psi(\varphi, J), I(\varphi, J)) + \frac{\partial}{\partial t} F_2(\psi(\varphi, J), J, t) \qquad (4.11)$$

As explained above, we seek a transformation that makes φ cyclic so that $K = K(J)$. The appropriate generating function does not depend on time, so (4.11) becomes

$$K(J) = H(\psi(\varphi, J), I(\varphi, J)) \qquad (4.12)$$

and the new equations of motion are

$$\dot{\varphi} = \frac{\partial K}{\partial J} \qquad \dot{J} = -\frac{\partial K}{\partial \varphi} \tag{4.13}$$

The approximation procedure consists in expanding the left and right sides of (4.12) in powers of ϵ and then equating terms of zeroth and first order. This procedure could be carried out to higher order. I'm interested in first order corrections only.

At first sight, this agenda looks hopeless. We need to know the *exact* value of J to make use any of these terms, even the zeroth order approximation. The exquisite point is that we can use (4.9) to calculate J exactly without knowing the complete transformation.

The zeroth order Hamiltonian is expanded with the help of (4.5).

$$H_0(I) = H_0\left(\frac{\partial F}{\partial \psi}\right) = H_0\left(J + \epsilon \frac{\partial F_1}{\partial \psi} + \cdots\right) = H_0(J) + \epsilon \frac{\partial F_1}{\partial \psi} \frac{\partial H_0}{\partial J}\bigg|_{\epsilon=0} + \cdots$$

$$\frac{\partial H_0(J)}{\partial J}\bigg|_{\epsilon=0} = \frac{\partial H_0(I)}{\partial I} = \omega \tag{4.14}$$

The first order term is already multiplied by ϵ.

$$\epsilon H_1(\psi, I) = \epsilon H_1(\psi, J) + \cdots$$

Now expand both sides of (4.12). We have, correct to first order in ϵ,

$$K_0(J) + \epsilon K_1(J) = H_0(I(\varphi, J)) + \epsilon H_1(\psi, I)$$

$$= H_0(J) + \epsilon \frac{\partial F_1}{\partial \psi} \omega + \epsilon H_1(\psi, J)$$

Equating equal powers of ϵ gives

$$K_0(J) = H_0(J) \tag{4.15}$$

$$K_1(J) = \frac{\partial}{\partial \psi} F_1(\psi, J)\, \omega + H_1(\psi, J) \tag{4.16}$$

The notation $H_0(J)$ means that you take your formula for $H_0(I)$ and replace the *symbol I* with the *symbol J without making any change in the functional form of H_0*.

Integrate (4.16) around one cycle and use (4.10); (4.16) becomes

$$K_1(J) = \overline{H_1}(J) \equiv \frac{1}{2\pi} \int_0^{2\pi} H_1(\psi, J)\, d\psi, \tag{4.17}$$

and

$$\frac{\partial}{\partial \psi} F_1(\psi, J) = \frac{1}{\omega(J)} \left[\overline{H_1} - H_1(\psi, J)\right] \equiv -\frac{\tilde{H}_1(\psi, J)}{\omega} \tag{4.18}$$

$\tilde{H}_1$ is the periodic part of H_1. We are left with a differential equation that is easy to integrate.

$$F_1(\psi, J) = -\frac{1}{\omega(J)} \int d\psi\, \tilde{H}_1(\psi, J) \tag{4.19}$$

Differentiating with respect to J gives the first-order correction to φ.

4.2 Summary

I will summarize all these technical details in the form of an algorithm for doing first order perturbation theory. Remember that the object is to find equations of motion in the form $q = q(t)$ and $p = p(t)$. We do this in three steps: (1) Find $q = q(\psi, I)$ and $p = p(\psi, I)$. (2) Find $I = I(\varphi, J)$ and $\psi = \psi(\varphi, J)$. (3) J and $\dot{\varphi}$ are constant, so $\varphi = \dot{\varphi}t + \varphi_0$.

(1) Identify the H_0 part of the Hamiltonian. Find the transformation equations $q = q(\psi, I)$ and $p = p(\psi, I)$ using the Hamiltonian-Jacobi equation as described in the previous chapter. Use (4.14) to get ω.
(2) Equation (4.9) can be used to find J in terms of the total energy E. The integral presents no difficulties in principle, especially if the Hamiltonian is separable. In fact, textbooks never bother to do this. It seems sufficient to display the results in terms of J, the assumption being that we could find $J = J(E)$ if we really had to.
(3) The first-order correction to the energy is obtained from the integral in (4.17). Get the first order correction to the frequency by differentiating it with respect to J.
(4) The generating function F_1 is calculated from (4.19). It is then substituted into (4.5) and (4.6). These give implicit equations for $\psi = \psi(\varphi, J)$ and $I = I(\varphi, J)$. Unfortunately, it is usually impossible to invert them to obtain these formula explicitly.

4.2.1 The simple pendulum

The pendulum makes a nice example

$$H = \frac{l^2}{2mR^2} + mgR(1 - \cos\theta)$$

The angular momentum $l = mR^2\dot{\theta}$ is canonically conjugate to the angle θ.

$$H = \frac{1}{2mR^2}\left[l^2 + m^2R^4\omega^2\theta^2\left(1 - \frac{\theta^2}{12} + \cdots\right)\right]$$

The first two terms reduce to the familiar harmonic oscillator with $\omega_0^2 = g/R$. This is the zeroth order problem.

$$H_0 = E_0 = \frac{l^2}{2mR^2} + \frac{mgR\theta^2}{2}$$

$$l^2 = \left(\frac{dW}{d\theta}\right)^2 = 2mR^2E_0 - m^2R^4\omega^2\theta^2$$

The last line makes use of (2.49). Make the natural substitution

$$\sin^2 \psi = \frac{mR^2\omega^2}{2E_0} \theta^2 \tag{4.20}$$

$$l = \left(\frac{dW}{d\theta}\right) = \sqrt{2mR^2E_0} \cos\psi \tag{4.21}$$

We can look on this as a convenient change of variable, but ψ is also the angle variable. This can be seen from Equation (2.46).

$$I = \frac{1}{2\pi} \oint l \, d\theta = \frac{\sqrt{2mR^2E_0}}{2\pi} \oint \left[1 - \frac{mR^2\omega^2}{2E_0} \theta^2\right]^{1/2} d\theta$$

Use (4.21) to get the familiar result, $I = E_0/\omega$. If we had not made the substitution $\omega^2 = g/R$ to start with we would have discovered this relationship at this point. The generating function is obtained from the indefinite integral

$$W = \int \left(\frac{dW}{d\theta}\right) d\theta = \int \left[2mR^2\omega I - m^2R^4\omega^2\theta^2\right]^{1/2} d\theta \tag{4.22}$$

According to the basic transformation formula we should have

$$\psi = \frac{\partial W}{\partial I}$$

One can show by differentiating (4.22) and using (4.21) to complete the integration, that this is indeed so.

Equations (4.20) and (4.21) can be rearranged to give

$$l = \sqrt{2mR^2I\omega} \cos\psi \tag{4.23}$$

$$\theta = \sqrt{\frac{2I}{mR^2\omega}} \sin\psi \tag{4.24}$$

The goal of the action-angle program is to express the original coordinates and momenta in terms of the action-angle variables. This has now been completed to zeroth order.

The first order correction is

$$H_1(\psi, I) = -\frac{mR^2\omega^2\theta^4}{24} = -\frac{I^2}{6mR^2} \sin^4\psi.$$

We are now in a position to recast our Hamiltonian à la (4.1).

$$H(\psi, I) = I\omega + \epsilon \left(-\frac{I^2}{6mR^2} \sin^4\psi\right)$$

We have also obtained $\omega = \sqrt{g/R}$ "for free." The ϵ is there for bookkeeping purposes only. We have no further need for it.

$$K_0(J) = H_0(J) = J\omega$$

$$K_1(J) = \overline{H_1(J)} = \frac{1}{2\pi} \int_0^{2\pi} H_1 \, d\psi = -\frac{J^2}{16mR^2}$$

$$\tilde{H} = H_1 - \overline{H_1} = \frac{J^2}{48mR^2}(3 - 8\sin^4 \psi)$$

$$F_1(\psi, J) = -\frac{1}{\omega} \int d\psi \, \tilde{H}_1 = \frac{J^2}{192 \, mR^2\omega}(\sin 4\psi - 8\sin 2\psi)$$

$$\omega = \frac{\partial K}{\partial J} = \omega - \frac{J}{8mR^2}$$

$$I = \frac{\partial F}{\partial \psi} = -\frac{\tilde{H}_1(\psi, J)}{\omega} = \frac{J^2}{48mR^2\omega}(8\sin^4 \psi - 3)$$

$$\varphi = \frac{\partial F}{\partial J} = \frac{J}{86mR^2\omega}(\sin 4\psi - 8\sin 2\psi)$$

We would like to disentangle these last two equations to get $\psi = \psi(\varphi, J)$ and $I = I(\varphi, J)$, but alas, it seems not quite possible.

4.3 Many degrees of freedom

For systems of two or more degrees of freedom, canonical perturbation theory is formulated in exactly the same way as before – but now profound difficulties arise, even to first order in ϵ. The problem centers around equation (4.18) repeated here for reference

$$\omega(J)\frac{\partial F_1(\psi, J)}{\partial \psi} = -\tilde{H}_1(\psi, J)$$

We were able to solve this with a simple integration (4.19). This is not possible for more that one degree of freedom, so we must resort to Fourier series. Before doing this, however, we will need to generalize our notation. Let's use the vectors

$$\mathbf{J} = (J_1, \ldots, J_n) \qquad \boldsymbol{\omega} = (\omega_1, \ldots, \omega_n) \qquad \boldsymbol{\nabla}_\psi = (\frac{\partial}{\partial \psi_1}, \ldots, \frac{\partial}{\partial \psi_n})$$

where n is the number of degrees of freedom. In this notation (4.18) becomes

$$\boldsymbol{\omega}(\boldsymbol{J}) \cdot \boldsymbol{\nabla}_\psi F_1(\boldsymbol{\psi}, \boldsymbol{J}) = -\tilde{H}_1(\boldsymbol{J}, \boldsymbol{\psi}) \tag{4.25}$$

where

$$\overline{H}_1(\boldsymbol{J}) = \frac{1}{(2\pi)^n} \int_0^{2\pi} d\psi_1 \cdots \int_0^{2\pi} d\psi_n H_1(\boldsymbol{J}, \boldsymbol{\psi}) \tag{4.26}$$

and $\tilde{H}_1 = H_1 - \bar{H}_1$. Since both sides of (4.25) are periodic, we can solve them with Fourier series.

$$\tilde{H}_1(\boldsymbol{J}, \boldsymbol{\psi}) = \sum_k A_k(\boldsymbol{J}) e^{ik \cdot \psi} \tag{4.27}$$

$$F_1(\boldsymbol{J}, \boldsymbol{\psi}) = \sum_k B_k(\boldsymbol{J}) e^{ik \cdot \psi} \tag{4.28}$$

where k is a vector of integers

$$\boldsymbol{k} = k_1, \ldots, k_n$$

It seems as if we could proceed as follows: $\tilde{H}_1$ is known at this point, so we can find A_k Substitute these definitions into (4.25) we get

$$B_k = i \frac{A_k}{\boldsymbol{\omega} \cdot \boldsymbol{k}} \tag{4.29}$$

Now here's the infamous problem. Suppose, for example, there were only two degrees of freedom. In this case the denominator of (4.29) would be

$$\boldsymbol{\omega} \cdot \boldsymbol{k} = \omega_1 k_1 + \omega_2 k_2 \tag{4.30}$$

You can see that if the winding number ω_1/ω_2 is a rational number, then for some k, B_k will be infinite. It seems that the slightest perturbation will blow this system into outer space! Even if the winding number is not rational, there will always be values of k that will make $\boldsymbol{\omega} \cdot \boldsymbol{k}$ arbitrarily small.

This problem was discovered in the early twentieth century, and all the effort of the most eminent mathematicians of the day failed to solve it. One opinion held that the slightest perturbation would cause the system to become "ergodic," that is to say, the trajectories would fill up all of phase space. Numerical calculations later showed that this was often not the case. Trajectories will often "lock in" to stable patterns. This has been the subject of much contemporary research. When and why do trajectories lock in, and what happens when they do not? The question of what trajectories remain stable under small perturbations is at least partly answered by the so-called KAM (Kolmogorov, Arnold, Moser) theorem. In the general case there is, if not a complete theory, at least a well-developed taxonomy. We will turn to these matters in the next chapter.

4.4 Problems

(1) The oscillator in Problem 2, Section 2.8 (Figure 2.1) could be represented with the Hamiltonian
$$H_0 = \frac{p^2}{2m} + V(q)$$
where $V = k|q|$.

   (a) Use perturbation theory to find the frequency as a function of energy to first order in ϵ for the Hamiltonian
   $$H = H_0 + \epsilon k|q|$$
   Compare your results with the exact solution.

   (b) Repeat the calculation with a perturbation
   $$H_1 = \epsilon q^2$$

(2) Consider a perturbed harmonic oscillator
$$H = \frac{p^2}{2} + \frac{\omega^2 x^2}{2} + \epsilon x$$

   (a) Find the frequencies to first order in ϵ

   (b) How do your results compare with the exact solution?

(3) The Duffing oscillator
$$\ddot{x} + \delta\dot{x} + \alpha x + \beta x^3 = \gamma\cos(\omega t)$$
is a well-studied example of a damped, driven, non-linear oscillator with lots of interesting chaotic behavior. A simpler system is given by
$$\ddot{x} + x + \epsilon x^3 = 0$$
It describes a harmonic oscillator with a non-linear restoring force $F(x) = -x(1 + \epsilon x^2)$.

   (a) Show that this equation can be derived from a Hamiltonian.

   (b) Find the frequencies to first order in ϵ.

The exact solution can be found using the method of quadratures. See Wikipedia for further details.

4.5 Sources and references

Goldstein, H., Poole, C., and Safko, J. (2002). *Classical Mechanics* (Addison-Wesley).

Matzner, R. A., and Shepley, L. C. (1991). *Classical Mechanics* (Prentice-Hall).

Tabor, M. (1998) *Chaos and Integrability in Nonlinear Dynamics* (John Wiley & Sons).

Chapter 5

Introduction to chaos

The canonical perturbation theory of the previous chapter is a lot of work, and in two or more degrees of freedom it summons up the ogre of small denominators. Several other forms of perturbation theory have been devised, but the problem of small denominators seems to be systemic. I will illustrate the limitations of perturbation theory by considering the van der Pol oscillator. This is a simple nonlinear, one-dimensional, second-order differential equation closely resembling a damped harmonic oscillator. It has stable solutions which can easily be found numerically, yet it has no known analytic solutions, and perturbation theory, on general principles, just can't work![1] We then go on to discuss linear stability theory. With these simple techniques you can analyze most nonlinear systems (the van der Pol oscillator is an exception) and get a qualitative picture of the phase space dynamics. In one degree of freedom (two-dimensional phase space) it will become immediately apparent where perturbation theory is possible and a qualitative idea of the motion of the system where it is not.

Higher dimensional spaces are not so easy to analyze, in part because they are hard to visualize and in part because they are often not integrable. It is this non-integrability that leads to chaos. Here we resort to the Poincaré section and the notion of discrete maps. The Poincaré-Birkhoff and KAM theorems can then tell us something about the onset and structure of chaos.

---

[1] It should be remembered that all the major developments in elementary particle theory over the last few decades starting with the standard model in the 1970's are based on the notion of spontaneous symmetry breaking. Spontaneous symmetry breaking, almost by definition, cannot be described with perturbation theory. When perturbation theory fails we always expect new physics. The same is true (to a lesser extent) in classical mechanics as well.

5.1   The total failure of perturbation theory

To get some feeling for how perturbation theory might be useless, look at the following "toy" example.

$$\ddot{x} = -x + \epsilon(x^2 + \dot{x}^2 - 1)\sin(\sqrt{2}t) \tag{5.1}$$

This looks like a harmonic oscillator with a resonant frequency $\omega = 1$ and a "small" driving term with a frequency $\omega = \sqrt{2}$. Obvious solutions are $x(t) = \sin t$ and $x(t) = \cos t$, which hold for all values of ϵ. If we set $\epsilon = 0$ then the solutions more generally are $x(t) = x_0\sin(t + t_0)$. This solution plotted on a phase space plot of $x(t)$ versus $\dot{x}(t)$ will be a circle with radius $r = x_0$. What would you expect for finite ϵ? There presumably are other solutions, but don't waste your time looking for them! You should convince yourself however, that there are *no* solutions of the form

$$x(t) = \sin t + \sum_{n=1}^{\infty} \epsilon^n f_n(t) \tag{5.2}$$

Also convince yourself that the trouble comes from the non-linear terms. The point is because of the nonlinearity, it is not possible to start with unperturbed solutions and get new solutions by adding to them.

A more interesting and oft-studied example is the van der Pol equation. It was first introduced by van der Pol in 1926 in a study of the nonlinear vacuum tube circuits of early radios.

$$\ddot{x} + \epsilon(x^2 - 1)\dot{x} + x = 0 \tag{5.3}$$

Again the $\epsilon = 0$ equations are $x(t) = x_0\sin(t + t_0)$. In phase space this is a circle of radius x_0. If we make ϵ ever so much larger than zero, however, something remarkable happens as shown in the first of the plots in Figure 5.1. Yes the orbit eventually becomes something like a circle, but *regardless of the initial conditions*, the radius $r \approx 2$. The same sort of behavior is shown in Figure 5.1 for larger values of ϵ. The shape of the final orbit is determined entirely by ϵ and is completely unaffected by the initial conditions. A curve of the sort is called a *limit cycle*. It's easy to see in vague way why the limit cycle exists. The term proportional to ϵ in (5.3) looks like an oscillator damping term, but its sign depends on whether x^2 is greater or less than 1. If it is greater, the oscillation is damped; if it is smaller the oscillation is "undamped." Indeed, if ϵ is made negative, the orbits either collapse to zero or diverge to infinity depending on the initial conditions. For obvious reasons the solutions with positive ϵ are said to be stable and those with negative ϵ are said to be unstable.

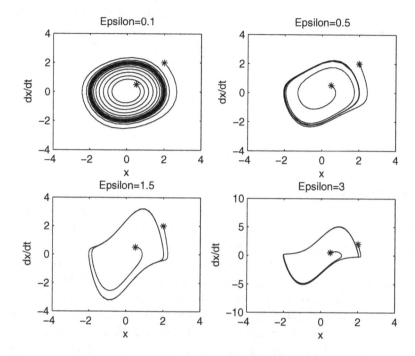

Fig. 5.1 The van der Pol plot for four values of ϵ and two starting values (indicated by asterisks).

This simple model makes an important point. Conventional perturbation theory starts with unperturbed, *i.e.* $\epsilon = 0$ solutions, and then looks for series solutions in powers of ϵ. This is obviously hopeless here since even a smidgeon of ϵ is enough to completely alter the nature of the orbits. It would be better to start with some simple function that approximated the limit cycle and then expand in powers of some parameter that characterized the deviation of the actual orbit from the simple function. Alas, I don't know how to do this. The trouble is that the limit cycle is so weird, at least for large ϵ, that it's hard to come up with a "lowest-order" solution. For many systems however, this is a practical approach. The trick is to look for the fixed points.

5.2 Fixed points and linearization

Equations of motion can always be cast in the form

$$\dot{\xi} = f(\xi, t) \tag{5.4}$$

With n degrees of freedom ξ and f are $2n$-dimensional vectors. For example, Hamilton's equations with one degree of freedom are

$$\dot{q} = \frac{\partial H}{\partial p}$$

$$\dot{p} = -\frac{\partial H}{\partial q}$$

$$\xi = \begin{bmatrix} p \\ q \end{bmatrix} \tag{5.5}$$

To keep the notation simple and general I will keep the notation in the form (5.4) for the time being and not type out the p's and q's. I will also restrict the discussion to *autonomous* systems, *i.e.* those in which the Hamiltonian does not depend explicitly on time.

A *fixed point* (also called a *stationary point, equilibrium point,* or *critical point*) is simply the point ξ_f where all the time derivatives vanish, $f(\xi_f, t) = \dot{\xi}_f = 0$. It's the place where nothing happens. In many cases the fixed points form the scaffolding on which the entire phase system is organized. Detailed information about the motion of a system close to a fixed point can be obtained by *linearizing* the equations of motion. This is done as follows. First, define ζ, the displacement vector

$$\zeta \equiv \xi - \xi_f \tag{5.6}$$

which is assumed to be small. Second, f is expanded in a Taylor series in powers of ζ.

$$\dot{\zeta}^j = \left. \frac{df^j}{d\zeta^k} \right|_{\xi_f} \zeta^k + O(\zeta^2) \equiv A_k^j \zeta^k + O(\zeta^2) \tag{5.7}$$

I am using the Einstein summation convention in which one sums over repeated indices. Dropping the $O(\zeta^2)$ terms gives the matrix equation

$$\dot{\zeta} = A \cdot \zeta \tag{5.8}$$

A is a constant matrix called (among other things) the *stability matrix*.

This would be a good point to review the properties of such matrices. If A is a real $m \times m$ matrix the following statements are true.

- A is diagonalizable if and only if all its eigenvectors are independent.
- If all the eigenvalues are different, then all the eigenvectors are independent, although not necessarily orthogonal.

- If some of the eigenvalues are equal, then A can be put in Jordan canonical form. Alternatively, one can use additional generalized eigenvectors to span the space.
- If the eigenvalues are complex, they always come in complex-conjugate pairs, λ and λ^*. The corresponding eigenvectors are independent.

Although this formalism is quite general, it's impossible to draw pictures in four or more dimensions, and so for purposes of illustration, I will restrict the discussion to systems with one degree of freedom. For the time being let's ignore the special case in which A has degenerate eigenvalues. The eigenvalues λ_1 and λ_2 are just the roots of the equation

$$\det |A - \lambda I| = 0 \tag{5.9}$$

The corresponding eigenvectors are D_1 and D_2 or in the case of complex eigenvalues, D and D^*.

The case in which the eigenvalues are real is particularly simple. The general solution to (5.8) is

$$\zeta(t) = c_1 D_1 e^{\lambda_1 t} + c_2 D_2 e^{\lambda_2 t} \tag{5.10}$$

where c_1 and c_2 are real constants determined by the initial conditions. If both eigenvalues are positive, all trajectories flow away from the fixed point. If they are both negative all trajectories flow toward the fixed point as in Figure 5.2(a). These are called *unstable* and *stable* fixed points respectively. If the eigenvalues have opposite signs then the trajectories are repelled from one axis and attracted to the other. This is called a *hyperbolic* fixed point or a *saddle point* as in Figure 5.2(b).

Suppose that A has degenerate eigenvalues. The corresponding canonical form is

$$A = \begin{vmatrix} \lambda & 1 \\ 0 & \lambda \end{vmatrix} \tag{5.11}$$

In terms of eigenvectors and generalized eigenvectors the solution is

$$\zeta = [c_1 D_1 + c_2 (D_2 + D_2 t)]e^{\lambda t} \tag{5.12}$$

As expected, the fixed point is stable or unstable depending on whether λ is negative or positive. An example of a stable point is shown in Figure 5.2(c).

Complex eigenvalues require a bit more discussion. The problem is that the eigenvectors, D and D^* are also complex. Of course one can define real vectors

$$D_1 = \frac{D + D^*}{2} \qquad D_2 = \frac{D - D^*}{2i} \tag{5.13}$$

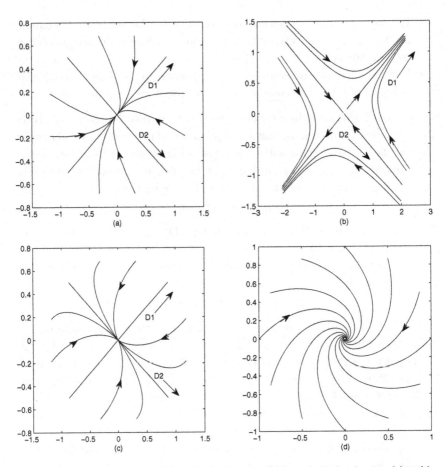

Fig. 5.2　Various fixed points: (a) stable fixed point, (b) hyperbolic fixed point, (c) stable fixed point with degenerate eigenvalues, (d) stable fixed point with complex eigenvalues.

so that $\boldsymbol{D} = \boldsymbol{D}_1 + i\boldsymbol{D}_2$, but they are not *eigen*vectors of anything, and this leads to an interesting complication. Consider the following. The formal solution of (5.8) is

$$\zeta = e^{\boldsymbol{A}t}\zeta_0 \tag{5.14}$$

where

$$e^{\boldsymbol{A}t} = \sum_{n=0}^{\infty} \frac{1}{n!}\boldsymbol{A}^n t^n \tag{5.15}$$

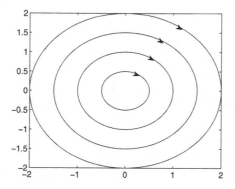

Fig. 5.3   An elliptic fixed point.

The operation of the exponential operator on one of the complex eigenvectors can be seen as follows. If $\lambda = \alpha + i\beta$ then

$$e^{\boldsymbol{A}t}\boldsymbol{D} = e^{\lambda t}\boldsymbol{D} = e^{\alpha t}(\cos\beta t + i\sin\beta t)(\boldsymbol{D}_1 + i\boldsymbol{D}_2) \qquad (5.16)$$

$$= e^{\alpha t}[\boldsymbol{D}_1\cos\beta t - \boldsymbol{D}_2\sin\beta t + i(\boldsymbol{D}_1\sin\beta t + \boldsymbol{D}_2\cos\beta t)]$$

$$= e^{\boldsymbol{A}t}\boldsymbol{D}_1 + ie^{\boldsymbol{A}t}\boldsymbol{D}_2$$

Separating real and imaginary parts gives

$$e^{\boldsymbol{A}t}\boldsymbol{D}_1 = e^{\alpha t}(\boldsymbol{D}_1\cos\beta t - \boldsymbol{D}_2\sin\beta t) \qquad (5.17)$$

$$e^{\boldsymbol{A}t}\boldsymbol{D}_2 = e^{\alpha t}(\boldsymbol{D}_1\sin\beta t + \boldsymbol{D}_2\cos\beta t)$$

Now let $\boldsymbol{\zeta}$ be a solution of (5.8). It can be expanded as follows

$$\boldsymbol{\zeta} = c_1(t)\boldsymbol{D}_1 + c_2(t)\boldsymbol{D}_2 \qquad (5.18)$$

where c_1 and c_2 are real time-dependent coefficients. Substituting (5.18) into (5.14) and using (5.17) gives

$$\boldsymbol{\zeta} = e^{\alpha t}[\boldsymbol{D}_1(c_{10}\cos\beta t + c_{20}\sin\beta t) + \boldsymbol{D}_2(-c_{10}\sin\beta t + c_{20}\cos\beta t)]\boldsymbol{\zeta}_0 \quad (5.19)$$

Let's concentrate on the coefficients. We will construct a coefficient vector

$$\boldsymbol{C} = \begin{bmatrix} c_1 \\ c_2 \end{bmatrix} \qquad (5.20)$$

Rewriting (5.19) in terms of coefficients gives

$$\boldsymbol{C}(t) = e^{\alpha t}\boldsymbol{R} \cdot \boldsymbol{C}_0 \qquad (5.21)$$

where R is just the rotation matrix

$$R = \begin{bmatrix} \cos \beta t & \sin \beta t \\ -\sin \beta t & \cos \beta t \end{bmatrix} \qquad (5.22)$$

The character of the solutions can be read off of (5.21). They are spirals that spiral in or out depending on whether α is positive or negative and spiral counterclockwise or clockwise depending on whether β is positive or negative. An example is shown in Figure 5.2(d). In the special case in which $\alpha = 0$, the trajectories are circles or ellipses depending on the initial conditions as shown in Figure 5.3.

It should be remembered that (5.8) is a linearized equation. It holds in some small region of the fixed point and of course as is so often the case, the theory gives us no way to tell how small that region might be. The following section analyzes two good examples of the method, and there are many other examples in the textbooks. On the other hand, the the theory fails completely for the van der Pol oscillator in the previous section.

5.2.1 *Two examples*

The following two examples are discussed in [Tabor (1989)]. Equations (5.23) and (5.24) are a simple mathematical model for what is jokingly called the lynx-rabbit cycle. The lynxes represented by y need to eat rabbits, but not too many or the rabbit population will collapse. The rabbits are represented by x, and although they may not agree with this, the rabbits need to be eaten or else there will be a population explosion and the food supply will be decimated. These two pressures keep in balance (or not) depending on the initial conditions.

$$\dot{x} = x - xy \qquad (5.23)$$

$$\dot{y} = -y + xy \qquad (5.24)$$

The system has two fixed points $(x_1, y_1) = (0, 0)$ and $(x_2, y_2) = (1, 1)$. The linearized equation is

$$\frac{d}{dt}\begin{bmatrix} \delta x \\ \delta y \end{bmatrix} = \begin{bmatrix} 1 - y_i & -x_i \\ y_i & -1 + x_i \end{bmatrix} \begin{bmatrix} \delta x \\ \delta y \end{bmatrix} \qquad (5.25)$$

where δx and δy are the displacements from the fixed points, and the subscript i in the array refers to the two fixed points. At $(x_1, y_1) = (0, 0)$ the eigenvalues are $\lambda = \pm 1$. The corresponding eigenvectors are orthogonal so the solution is

$$\begin{bmatrix} \delta x \\ \delta y \end{bmatrix} = c_1 \begin{bmatrix} 1 \\ 0 \end{bmatrix} e^{+t} + c_2 \begin{bmatrix} 0 \\ 1 \end{bmatrix} e^{-t} \qquad (5.26)$$

Evidentally there is a saddle point at the origin. At $(x_2, y_2) = (1,1)$ the eigenvalues are $\lambda = \pm i$. This is the signature of an elliptic point. The resulting phase plane is shown in Figure 5.4. The initial conditions were chosen at random and Equations (5.23) and (5.24) were solved numerically. Our simple linear approach gets everything just right.

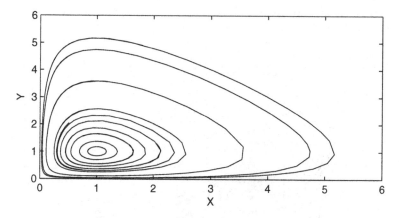

Fig. 5.4 Solutions for Equations (5.23) and (5.24). Initial conditions were chosen randomly.

The following two equations

$$\dot{x} = x(4 - x - y) \tag{5.27}$$

$$\dot{y} = y(x - 2) \tag{5.28}$$

have three fixed points $(x_1, y_1) = (0,0)$, $(x_2, y_2) = (1,1)$, and $(x_3, y_3) = (4,0)$ and the linearized equations

$$\frac{d}{dt} \begin{bmatrix} \delta x \\ \delta y \end{bmatrix} = \begin{bmatrix} 4 - 2x_i - y_i & -x_i \\ y_i & x_i - 2 \end{bmatrix} \begin{bmatrix} \delta x \\ \delta y \end{bmatrix} \tag{5.29}$$

where $i = 1, 2, 3$.

At $(x_1, x_2) = (0,0)$ the $\boldsymbol{A}$ matrix is diagonal and the eigenvalues are obviously $\lambda_{1,2} = -2, 4$. This is a hyperbolic fixed point with eigenvectors along the x and y axes. At $(x_2, y_2) = (2,2)$ the eigenvalues are $\lambda_{\pm} = -1 \pm i\sqrt{3}$, so this is a stable spiral point. Finally at $(x_3, y_3) = (4,0)$ the eigenvalues are $\lambda_{1,2} = (-4, 2)$. This is a hyperbolic fixed point but the eigenvectors are not orthogonal. The general solution is

$$\begin{bmatrix} \delta x \\ \delta y \end{bmatrix} = c_1 \begin{bmatrix} 1 \\ 0 \end{bmatrix} e^{-4t} + c_2 \begin{bmatrix} -2 \\ 3 \end{bmatrix} e^{2t} \tag{5.30}$$

The solutions for (5.27) and (5.28) with many randomly chosen initial conditions are shown in Figure 5.5. The landscape is dominated by the stable point at $(2, 2)$ which draws all the trajectories into itself. The effect of the fixed point at $(0, 0)$ is to draw all the trajectories down along the y direction and push them away in the x direction. Its effect is felt at long distance. The effect of the fixed point at $(4, 0)$ is barely visible. Our fixed point analysis has no way of gauging the relative strengths of the various points.

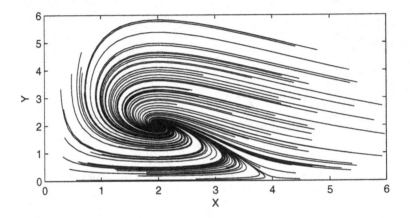

Fig. 5.5   Solutions for Equations (5.27) and (5.28).

5.3   The Henon-Heiles oscillator

Although the theory from the previous section is perfectly general in the sense that it can be applied to systems with any number of degrees of freedom, it is almost impossible to visualize in four or more dimensions, and the number of cases that must be considered increases rapidly. The best tool for visualizing higher dimensional spaces is the Poincaré section, which was described briefly in Chapter 3. A good example of motion with two degrees of freedom is the fascinating and oft-studied case of the Hénon-Heiles Hamiltonian. The Hamiltonian was originally used to model the motion of stars in the galaxy.[2] Written in terms of dimensionless variables the Hamiltonian is

$$H = \frac{1}{2}(\dot{x}^2 + \dot{y}^2 + k_1 x^2 + k_2 y^2) + \lambda(x^2 y - \frac{y^3}{3}) \qquad (5.31)$$

---

[2]See [Goldstein *et al.* (2002)] for a review of the physics.

This is the Hamiltonian of two uncoupled harmonic oscillators with a perturbation proportional to λ. The unperturbed part of the Hamiltonian is just Equation (3.40) with different notation. The oscillators have frequencies $\omega_1 = \sqrt{k_1}$ and $\omega_2 = \sqrt{k_2}$. The phase space is the four-dimensional space spanned by x, $\dot{x}$, y, and $\dot{y}$. We can think of the unperturbed orbits as lying on a torus; the lines of constant x and $\dot{x}$ are lines of "latitude" and lines of constant y and $\dot{y}$ are lines of "longitude" (or the other way around). In either case, lines of latitude and lines of longitude are circles.

It will turn out as we go along that the winding number is very important. Just for the record let's define

$$w \equiv \frac{\omega_1}{\omega_2} \tag{5.32}$$

If the winding number is rational, $w = r/s$, x will complete r cycles while y completes s. Let us make a Poincaré section through the $x = 0$ plane. Each time an orbit passes through $x = 0$ with $\dot{x} > 0$ we mark a point at y and $\dot{y}$ on the $x = 0$ plane. An example is shown in Figure 5.6 for $w = 7/2$. Because $r = 7$ there are seven discrete dots on the Poincaré section. The orbit crosses the $x = 0$ plane in the order labeled. Because $s = 2$ the orbit has to go around the seven-sided figure twice to visit all the points. You should convince yourself that there are six possible itineraries corresponding to the possible values of s between one and six. The case of an irrational winding number is shown in Figure 5.7. The x vs. y plot is completely filled in, *i.e.* the trajectory visits every point in phase space, and the Poincaré plot is a continuous loop. Continuous loops like this on the Poincaré plot are a sign that the system is circulating around an invariant torus. If this is true for all possible initial conditions, the system is integrable.

When we turn on the perturbation by making $\lambda \neq 0$ something remarkable happens. I will illustrate this by setting $k_1 = k_2 = \lambda = 1$ and changing the amount of perturbation by changing the total energy through (5.31) with $x = 0$. The initial conditions are specified by choosing y and $\dot{y}$ and fixing $\dot{x}$ through

$$\dot{x} = \left(2E - \dot{y}^2 - y^2 + \frac{2}{3}y^3\right)^{1/2} \tag{5.33}$$

We start with $E = 1/12$. Figure 5.8 shows phase space trajectories produced by several different initial conditions. The trajectories seem to be circulating around several smooth but distorted tori. Figure 5.9 with $E = 1/8$ appears to show five separate trajectories corresponding to five different initial conditions. In fact, this is a Poincaré plot of a single trajectory

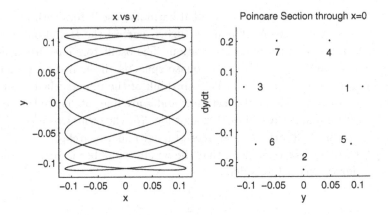

Fig. 5.6   Two uncoupled harmonic oscillators with $w = 7/2$.

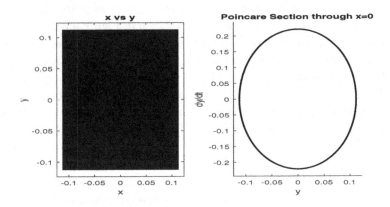

Fig. 5.7   Same as Figure 5.6 but with an irrational winding number.

that loops back and forth apparently randomly through the $x = 0$ plane, but in just such a way as to mark out the five little "islands." With $E = 1/6$ the trajectory's path becomes completely random and the Poincaré plot consists of random dots. The five islands from Figure 5.9 are superimposed for comparison.

Figures 5.6 through 5.10 show a progression from the orderly motion of the uncoupled oscillators to complete chaos. Figure 5.8 suggests orderly motion around several distorted tori. As the interaction strength increases each trajectory traces out isolated patches. The points then begin to disperse in a random way with some structure remaining. In the last plot the

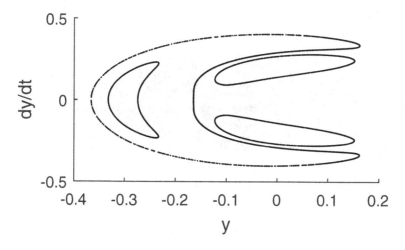

Fig. 5.8   Smooth orbits with $E = 1/12$.

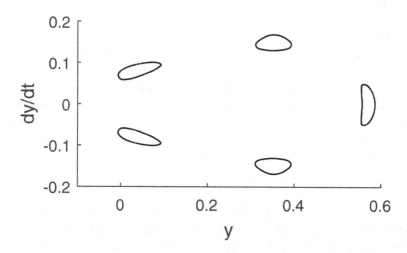

Fig. 5.9 $E = 1/8$, A single orbit crosses the $x = 0$ plane at five small regions.

points are arranged in a completely random pattern. This is paradigmatic. As the strength of the perturbation increases orderly motion disintegrates into chaos. One of the goals of chaos theory is to explain and predict this phenomena. This will require some new formalism.

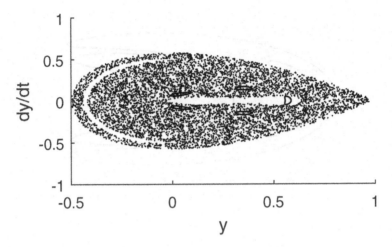

Fig. 5.10 $E = 1/6$, Complete chaos. Note the islands superimposed from the previous figure.

5.4   Discrete maps

Poincaré plots consist of a sequence of points. The coordinates of the points are canonically conjugate variables. In the example of the Henon-Heiles oscillator the coordinates were x and $\dot{x}$, but in other instances they might be action-angle variables or any other conjugate pair. Since I will be discussing general properties of these plots, I will simply call the variables x and y. The most important thing about these points is that they form a definite sequence, $(x_1, y_2), (x_2, y_2), \cdots, (x_n, y_n)$, and this implies a mapping operator T that maps the n'th point into the $(n + 1)$'th point.

$$\mathrm{T}(x_n, y_n) = (x_{n+1}, y_{n+1}) \tag{5.34}$$

We assume that for every (x_n, y_n) there is a unique (x_{n+1}, y_{n+1}) and that T is invertible. In general we will not be able to find an explicit formula for T. Imagine, for example, coming up with such an operator for the Henon-Heiles oscillator. In fact, the utility of this approach is that it enables us to analyze non-integrable systems for which exact analytic solutions, and with them any explicit formulas for T, are impossible almost by definition.

The second most important thing about these points is that the transformation $(x_n, y_n) \rightarrow (x_{n+1}, y_{n+1})$ is a canonical transformation. This is a consequence of the fact that Hamiltonian flow is itself a canonical transformation. There are numerous "toy" models with explicit formulas for T.

We can check to make sure that they are canonical. They must satisfy the following identity from Theorem 3.2.

$$\frac{(\partial x_{n+1}, \partial y_{n+1})}{(\partial x_n, \partial y_n)} = 1 \tag{5.35}$$

One such toy model is called the *standard map*, presumably because it appears in so many different contexts.

$$\{\phi_{n+1} = \phi_n + J_{n+1}\} \bmod 2\pi \tag{5.36}$$

$$\{J_{n+1} = \epsilon \sin \phi_n + J_n\} \bmod 2\pi$$

[Jose and Saletan (1998)] derive it from a system called the "kicked rotator" in which a periodic impulsive force is applied to a rotating object. The system lends itself easily to action-angle variables; ϕ is the angle evidentally, and J is the angular momentum. Both change discontinuously at each "kick"; ϕ_n and J_n are the coordinates after the n'th kick. This meets the above requirements. Equation (5.35) is satisfied and the transformation can be inverted algebraically. Despite the highly contrived nature of this transformation, we will see that it has all the features of more realistic systems. The reasons for this will be explored in subsequent sections.

Not only ϕ, but also J is periodic with period 2π. We can imagine the points as lying along orbits wrapped around a cylinder, or equivalently, lying on the surface of a one-dimensional torus. If x_n lies on one orbit, then x_{n+1} is guaranteed to be on the same orbit. If $\epsilon = 0$ the system is trivial and the orbits are straight lines. We can think of ϵ as a parameter that "turns on" a non-linear perturbation. The point of this exercise is to understand how the dynamics of the system change as ϵ is increased. In a word, the points break away from orbits and scatter around randomly; but the devil as they say, is in the details.

The $\epsilon = 0$ case is shown in Figure 5.11. For this and subsequent plots, the J_0's were chosen randomly and Equation (5.36) was iterated 1000 times. Notice that some of the orbits consist of discrete dots, some of dashes, and some of solid lines. The difference has to do with winding number. If $J = 2\pi r/s$, with r and s integers, there will be s discrete dots, and the sequence they follow will wrap around the cylinder r times.

When ϵ is increased to 0.050 a new feature appears, a loop in the center of the plot as shown in Figure 5.12. This is unusual in the sense that it can be contracted to a point; it is topologically distinct from all the $\epsilon = 0$ circles. It will turn out that mapping operators like (5.34) have fixed points like the continuous functions studied in Section 5.2, but with a much simpler

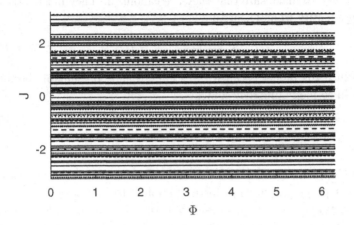

Fig. 5.11   The standard map with $\epsilon = 0$.

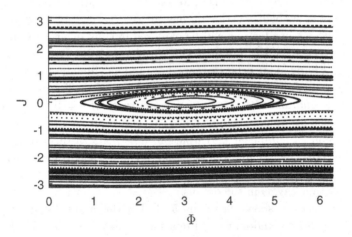

Fig. 5.12   The standard map with $\epsilon = 0.050$.

taxonomy. The center of the loop at $\phi = \pi$, $J = 0$ is an example of an elliptic or stable fixed point, and in fact, it's surrounded by an ellipse!

Look at Figure 5.13 with $\epsilon = 0.5$. The elliptic fixed point is still apparent, although there is a new feature at $\phi = 0$, $J = 0$. This is a hyperbolic or unstable fixed point, and in keeping with its name, there is a thin band of random points surrounding it, but the elliptic point seems to be radi-

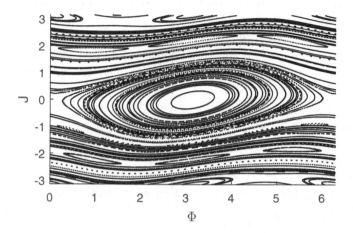

Fig. 5.13 The standard map with $\epsilon = 0.500$.

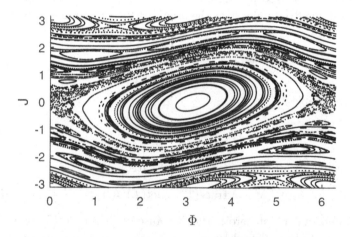

Fig. 5.14 The standard map with $\epsilon = 0.750$.

ating its stability in all directions. An assortment of smaller loops have appeared. It seems that the central elliptic point is spawning clones of itself, and although it's not apparent in this plot, they are separated by hyperbolic points that will generate more chaos as ϵ is increased.

In Figure 5.14 the situation has deteriorated. The random points are spreading. Intricate "necklaces" of stable ellipses have appeared, all fighting a losing battle against the spreading chaos. When ϵ is increased further the points become completely random.

The standard map seems trivial, but in fact most conservative, two-degrees-of-freedom systems generate Poincaré plots that look just like the ones above. The reasons will emerge after we have had a closer look at fixed points.

5.5   Linearized maps

Like the continuous transformations we studied in Section 5.2, discrete maps have fixed points about which one can analyze the local topology. A generic mapping like (5.34) can be written a bit more explicitly as

$$x_{n+1} = f(x_n, y_n) \qquad y_{n+1} = g(x_n, y_n) \tag{5.37}$$

A fixed point of the mapping would be a point where $x_{i+1} = x_i$ and $y_{i+1} = y_i$. I will argue later on that in a plot like Figure 5.14 there are an infinite number of fixed points, but to keep the algebra simple here I will assume that the fixed point is at the origin $(0,0)$. Linearizing T about this point gives

$$\begin{bmatrix} \delta x_{i+1} \\ \delta y_{i+1} \end{bmatrix} = \begin{bmatrix} T_{11} & T_{12} \\ T_{21} & T_{22} \end{bmatrix} \begin{bmatrix} \delta x_i \\ \delta y_i \end{bmatrix} \tag{5.38}$$

where for example

$$T_{11} = \left. \frac{\partial f(x,y)}{\partial x} \right|_{x,y=0} \tag{5.39}$$

The eigenvalues λ_i of the T_{ij} matrix must satisfy

$$\lambda^2 - \lambda \operatorname{trace}(T) + \det(T) = 0 \tag{5.40}$$

The all-important point here is that because of (5.35), $\det(T) = 1$. This greatly restricts the allowed types of fixed points. There are only three cases to consider.

If $|\operatorname{trace}(T)| < 2$, $\lambda_1 \lambda_2$ are a complex conjugate pair lying on the unit circle, that is,

$$\lambda_1 = e^{+i\alpha}, \qquad \lambda_2 = e^{-i\alpha} \tag{5.41}$$

This is simply a rotation in the vicinity of the fixed point $(0,0)$. This corresponds to a stable or elliptic point. Thus in the immediate neighborhood of $(0,0)$ we expect to find invariant curves that are circles or ellipses.

If $|\text{trace}(T)| > 2$, $\lambda_1\,\lambda_2$ are real numbers satisfying

$$\lambda_1 = 1/\lambda_2 \tag{5.42}$$

There are two subcases to consider here depending on whether λ is positive or negative. If it is positive we have a regular hyperbolic fixed point in which successive iterates stay on the same branch of the hyperbola as in Figure 5.15(a). If $\lambda < 0$ we have a hyperbolic-with-reflection fixed point in which successive iterates jump backwards and forwards between opposite branches of the hyperbola as in Figure 5.15(b).

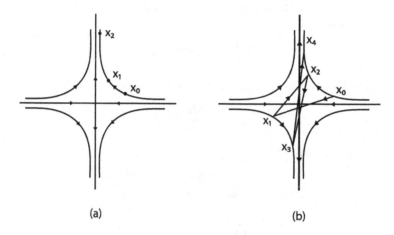

(a)   (b)

Fig. 5.15   (a) Hyperbolic fixed point. (b) Hyperbolic-with-reflection fixed point.

Except for the special case $\lambda = 1$ the eigenvectors are independent (though not necessarily orthogonal). We can also identify a locus of points all of which are mapped directly away from the fixed point by repeated applications of T. Since this amounts to multiplication by λ, *i.e.* $Tx = \lambda x$ with $\lambda > 1$, x is pushed asymptotically away from the fixed point by re-peated applications of T. This locus of points is called the *unstable manifold*. There is another locus of points corresponding to the eigenvalue $1/\lambda$ which are drawn asymptotically into the fixed point. This is the *stable manifold*. In the linear approximation these are straight lines lying along the eigenvectors, but even when the linear approximation fails spectacularly, the notion of stable and unstable manifolds is meaningful and important as we will see in Section 5.7.

5.6   Lyapunov exponents

Loosely speaking, systems are chaotic because adjacent trajectories diverge exponentially from one another. If this were literally true we could parameterize this divergence with the function $e^{\lambda x}$, where λ is some constant and x is the independent variable, which might be continuous or discrete depending on the application. This is the basic idea behind Lyapunov exponentials, a formalism with many alternate definitions (and spellings).

Let's apply this idea first to a one-dimensional iterative map of the form

$$x_{i+1} = f(x_i) \tag{5.43}$$

We can characterize the divergence of two trajectories separated by ϵ upon the n-th iteration as

$$\lim_{\epsilon \to 0} \frac{|f(x_n + \epsilon) - f(x_n)|}{\epsilon} = \left| \frac{df(x_n)}{dx_n} \right| \tag{5.44}$$

A small but finite deviation at the n-th iteration, say δx_n, should grow to

$$\delta x_{n+1} \approx \left| \frac{df(x_n)}{dx_n} \right| \delta x_n \tag{5.45}$$

Continuing this reasoning

$$\frac{\delta x_{n+1}}{\delta x_0} = \left| \frac{df(x_n)}{dx_n} \frac{df(x_{n-1})}{dx_{n-1}} \cdots \frac{df(x_0)}{dx_0} \right| \tag{5.46}$$

$$= \prod_{i=0}^{n} |f'(x_i)| = (?) \, e^{\lambda n}$$

The last equality can't be literally true. λ will certainly depend on the point n where we stop iterating. We should write instead

$$\lambda(n) = \frac{1}{n} \ln \prod_{i=0}^{n} |f'(x_i)| \tag{5.47}$$

with the understanding that the definition only makes sense if there is some range of n over which $\lambda(n)$ is more or less constant. λ defined in this way is a Lyapunov exponent.

In the case of multidimensional mappings, use the linearized map defined by (5.38), except not necessarily evaluated at a fixed point but rather any arbitrary $\boldsymbol{x}_i$. We will use the eigenvalues $\sigma_i(n), i = 1, \ldots, n$, of the matrix

$$\boldsymbol{T}(\boldsymbol{x}_n)\boldsymbol{T}(\boldsymbol{x}_{n-1}) \cdots \boldsymbol{T}(\boldsymbol{x}_1) \tag{5.48}$$

where $T(x_i)$ is the linearization of T at the point x_i, The Lyapunov exponents are defined as

$$\lambda_i(n) = \frac{1}{n} \ln |\sigma_i(n)| \tag{5.49}$$

Since the T's have unit determinant for area-preserving maps, it is clear that the sum of the exponents must be zero.

For the final example, suppose the equation of motion is

$$\dot{x} = f(x) \tag{5.50}$$

Let $s(t) = x(t) - x_0(t)$ be the difference between two near-by trajectories. If this does indeed diverge exponentially with time, then $\dot{s} = \lambda s$. Then we can argue that

$$\dot{s} = \dot{x} - \dot{x}_0 = f(x) - f(x_0) = \lambda s = \lambda(x - x_0) \tag{5.51}$$

$$\lambda = \frac{f(x) - f(x_0)}{x - x_0} \approx \left. \frac{df}{dx} \right|_{x_0} \tag{5.52}$$

5.7 The Poincaré-Birkhoff theorem

The phase-space trajectories of integrable systems move on smooth tori. The appearance of the Poincaré section depends on whether the winding number is rational or irrational. If it is rational the section shows discrete points. If irrational, the points form a continuous loop. Under the influence of nonlinear perturbations the tori become distorted, then break up into smaller tori, and finally disintegrate into chaos. It turns out that the way this happens depends on whether the winding number is rational or irrational. If it is irrational the tori are distorted but preserved under small perturbations. This is a gross oversimplification of the KAM theorem, which I will discuss in Sections 5.10 through 5.12. If the winding number is rational, the tori break up in a way that is governed by the so-called Poincaré-Birkhoff theorem, the subject of this section. This may seem like a swindle, since every irrational number can be approximated to arbitrary accuracy by a rational number. But as it turns out, some numbers are more irrational than others.

I will illustrate the PB theorem using the standard map Equation (5.36), but the theorem is quite general. It depends only on the general properties of discrete maps discussed at the beginning of Section 5.4. I will use the symbol T_ϵ as in (5.34), but with a subscript ϵ to indicate that some perturbation might or might not be present.

$$T_\epsilon(\phi_n, J_n) = (\phi_{n+1}, J_{n+1}). \tag{5.53}$$

Now imagine the points in Figure 5.11 ($\epsilon = 0$) plotted in polar coordinates (for positive J) with ϕ the angular and J the radial coordinate. The points now lie on concentric circles of constant J. Choose $J = J_r \equiv 2\pi j/k$, with k and j integers, *i.e.* J_r has a rational winding number. If we iterate T_0 k times, J remains unchanged and ϕ is incremented by j factors of 2π, which is to say ϕ is not changed at all. Symbolically

$$T_0^k(\phi, J_r) = (\phi, J_r)$$

Theorem 5.1. *A torus with rational winding number j/k is invariant under T_0^k, i.e. every point on the torus is a fixed point of T_0^k. When ϵ is even slightly larger than zero, only a discrete (even) number of fixed points of T_ϵ^k survive. They are alternately elliptic and hyperbolic.*

Proof. Take a J_+ slightly larger than J_r. T_0^k will increment ϕ by slightly more than $2\pi j$ so ϕ will increase. In the same way if $J_- < J_r$, T_0^k will cause ϕ to decrease. We can imagine the values of ϕ lying on three circles J_+, J_r, and J_- as shown in Figure 5.16(a). Put another way, the *flow* outside J_r is clockwise; inside it is counterclockwise.

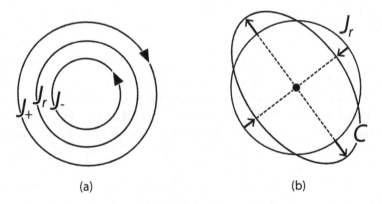

(a)   (b)

Fig. 5.16   (a) Three orbits of the unperturbed standard map T_0^k. (b) The ϕ coordinate is left invariant on C by the perturbed map T_ϵ^k.

Now turn on a small perturbation $\epsilon > 0$. T_ϵ^k will map some ϕ's to larger values and some to smaller, but there will be some locus of points, called C

in Figure 5.16(b) on which the ϕ's are not changed at all. In other words, the curve J_r is mapped purely radially into C.

$$T_\epsilon^k(J_r, \phi) = (J_C, \phi)$$

Curve C is mapped in turn into a new curve called D in Figure 5.17.

$$T_\epsilon^k(J_C, \phi) = (J_D, \phi')$$

The curves C and D must have the same area (remember these are area-preserving transformations) so they must cross one another an even number of times. This situation is shown in Figure 5.17. At the crossings however,

$$T_\epsilon^k(J_C, \phi) = (J_C, \phi)$$

in other words, the crossings represent points that are invariant under T_ϵ^k since they are fixed points.

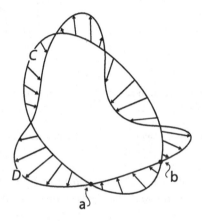

Fig. 5.17   The curves C and D. Crossings like a and b are fixed points.

You can ascertain the type of fixed points by seeing how other points in their immediate vicinity are mapped. Compare this *flow* as it's called with the arrows in Figures 5.2(b) or 5.15(a) for hyperbolic fixed points and Figure 5.3 for elliptic fixed points. You should be able to convince yourself that the points along the curve C are alternatively hyperbolic and elliptic. Figures 5.18 and 5.19 should help you visualize this. Since there are an even number of fixed points, half of them will be elliptic and half hyperbolic.

$\square$

This is clear in the first graph in Figure 5.20. With $k = 1$ there is a elliptic fixed point at $\phi = \pi$ and a hyperbolic fixed point at $\phi = 0$. All the

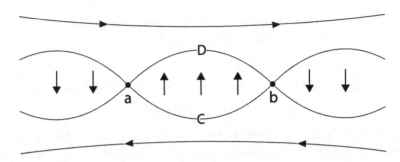

Fig. 5.18   A closer look at the fixed points a and b. The arrows indicate the mapping of T_ϵ^k. It is assumed that since ϵ is small the flow outside C and D (upper curve) is still clockwise like J_+ in Fig. 5.16(a) and the flow inside (lower curve) is counterclockwise.

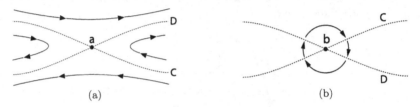

Fig. 5.19   (a) The flow in the immediate vicinity of a. (b) The flow in the immediate vicinity of b.

trajectories are trapped between these two points. The case with $k = 2$ is more complicated.

Theorem 5.2. *When $k > 1$ every fixed point of T_ϵ^k lying on the curve C, is accompanied by $k - 1$ additional fixed points of T_ϵ^k with the same "gender."*

Proof. Suppose (ϕ_0, J_0) is a fixed point of T_ϵ^k. We can create more fixed points by multiplying by T_ϵ as the following simple argument shows.

$$T_\epsilon^k \left[T_\epsilon(\phi_0, J_0) \right] = T_\epsilon T_\epsilon^k(\phi_0, J_0) = T_\epsilon(\phi_0, J_0)$$

Starting with (ϕ_0, J_0) we can create $k - 1$ additional fixed points by multiplying repeatedly with T_ϵ. To put it another way, every fixed point of T_ϵ^k is a member of a family of k fixed points obtained by multiplying by various powers of T_ϵ Because each mapping is a continuous function of ϕ and J, all the members of an elliptic family are elliptic and all the members of a hyperbolic family are hyperbolic. I claim that all the members of a family are distinct. Proof: Let (ϕ_s, J_s) be the fixed point obtained by $T_\epsilon^s(\phi_0, J_0) = (\phi_s, J_s)$ with $s < k$. Then of course all such points are fixed points of T_ϵ^k. The claim is that there is no $m < k$ such that

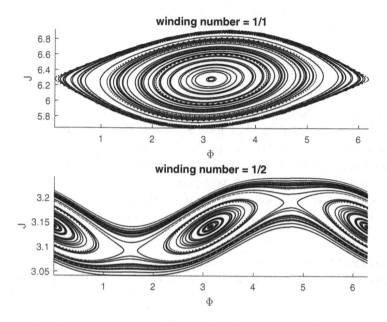

Fig. 5.20 Two plots of the standard map with $J_\epsilon^k = 2\pi j/k$ and $\epsilon = 0.1$. Each trajectory corresponds to a different starting value of ϕ.

$T_\epsilon^m(\phi_s, J_s) = (\phi_s, J_s)$. Multiply both sides of this equation with T_ϵ^{-s}. The result is $T_\epsilon^m(\phi_0, J_0) = (\phi_0, J_0)$ It is just this equation with m replaced by k that defines (ϕ_0, J_0). Hence $m = k$. This is a general result that applies to many dynamical systems with one degree of freedom. In its full generality it is known as the Poincaré-Birkhoff theorem. □

The second graph in Figure 5.20 makes sense in the light of this theorem. Since $k = 2$ there must be pairs of hyperbolic and elliptic fixed points. Each application of T_ϵ maps one of the fixed points into its partner.

All the points in Figure 5.20 seem to lie on smooth periodic curves called *orbits*. Each orbit is determined uniquely by a complete set of starting coordinates (ϕ_0, J_0). But in this regime it is J_0 that determines the "landscape" that the orbits see, *i.e.* it determines the fixed points and the way they are related to one another. As ϵ is increased this simple picture of fixed points and orbits breaks down. For example it is possible that one value of J can produce multiple orbits. Figure 5.20 shows two examples. Each graph is derived from a single value of J. As ϵ increases, there will also be regions in which the points seem to be distributed completely randomly. This is the onset of chaos.

5.8   All in a tangle

Have another look at the hyperbolic fixed points in Figures 5.2(b) and 5.15(a). There are always two loci of points leading directly toward the fixed point and two loci leading away from it. These are called the stable and unstable manifolds respectively. In the usual notation, a stable manifold is called H_+ and an unstable one, H_-. Call the fixed point P_f. Any point along H_+ will be mapped asymptotically back to P_f under repeated applications of T_ϵ, and any point on H_- will be mapped asymptotically back to P_f under repeated applications of T_ϵ^{-1}. Can these manifolds cross one another?

Theorem 5.3. *An unstable manifold cannot cross itself nor can it cross another unstable manifold emanating from another hyperbolic fixed point. The same is true of stable manifolds.*

Proof. At first sight this seems obvious. If there is a crossing at X then TX would map X into two different points, and this seems impossible. The flaw in this argument is that the manifold might cross itself a second time, and TX might lie at the second crossing. This is exactly what happens when an unstable manifold crosses another stable manifold. It cannot happen in the case of like manifolds as the following argument proves. Suppose an unstable manifold H_- of the fixed point P_f intersected itself at X. This situation is shown in Figure 5.21. Then H_- would contain a loop L from X back around to X again. Consider a sequence of points $a, b, c, \ldots$ lying on L with a closest to X and subsequent points "downstream." Successive mappings by T will produce a new sequence of points $T^k a, T^k b, T^k c, \ldots$ For sufficiently large k the points will have reversed their order. Those farthest from X (as measured along the loop) are now closest, and those closest are now farthest. This is impossible for a continuous operator. It's easy to construct similar arguments to prove the other three assertions of the theorem.   □

It is entirely possible for an unstable trajectory to loop around and become the stable trajectory of its own fixed point or to become the stable manifold of another fixed point. This is typical of stable integrable systems. The interesting situation, from the point of view of chaos, is when an

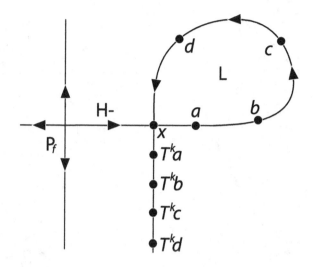

Fig. 5.21 An unstable manifold crossing itself.

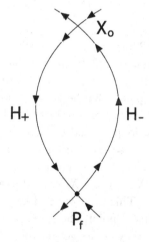

Fig. 5.22 A homoclinic intersection. The unstable manifold H_- crosses the stable manifold H_+ at X_0. Both manifolds emanate from the fixed point P_f.

unstable manifold *crosses* a stable manifold. It might cross the manifold emanating from its own fixed point as in Figure 5.22, in which case the crossing it is called a *homoclinic* point, or it might cross the stable manifold emanating from a different fixed point, in which case it is a *heteroclinic* point.

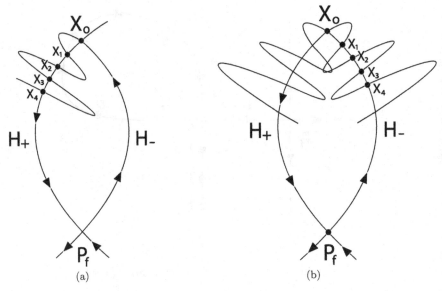

Fig. 5.23   (a) Adding the forward maps $\mathrm{T}^k X_0$. (b) Adding the backward maps $\mathrm{T}^{-k} X_0$.

Referring to Figure 5.23(a), X_0 is a homoclinic point that lies at the intersection of the unstable manifold H_- and stable manifold H_+ emanating from the fixed point P_f. Since both manifolds are invariant under T_ϵ, the $\mathrm{T}_\epsilon^k X_0$ are a set of discrete points that lie on both manifolds, so the two manifolds must therefore intersect again. For instance, because $X_1 = \mathrm{T}_\epsilon X_0$ is on both manifolds, H_- must loop around to meet H_+. Similarly, $X_k = \mathrm{T}_\epsilon^k X_0$ must lie on both manifolds, so H_- must loop around over and over again as illustrated in Figure 5.23(a). The inverse map also leaves H_+ and H_+ invariant and hence the $X_{-k} = \mathrm{T}_\epsilon^{-k} X_0$ are intersections that force H_+ to loop around to meet H_-. This is shown in Figure 5.23(b). As k increases and X_k approaches the fixed point, the spacing between the intersections gets smaller, so the loops they create get narrower. But because T_ϵ is area-preserving, the loop areas are the same, so the loops get longer, which leads to an arbitrarily large number of intersections among them. This situation is called a homoclinic tangle. Nonetheless the mess is contained, at least for small ϵ. Since stable manifolds cannot cross, the stable manifold emanating from P_f acts as a barrier to other stable manifolds. The same is true of the unstable manifolds.

Theorem 5.4. *Manifolds can't cross periodic orbits.*

Proof. If a manifold H crossed an orbit O at one point, it would have to cross an infinite number of times, but since O is periodic, successive applications of T would eventually cause H to cross itself and this we know to be impossible. □

As a consequence tangles develop in the vicinity of hyperbolic fixed points sandwiched on the left and right by orbits circulating around elliptic fixed points and bounded above and below by complete orbits circulating around the invariant torus.

5.9 An example: coupled pendulums

The relevance of the Poincaré-Birkhoff theorem goes way beyond the very artificial kicked rotator that the standard map is based on. The coupled pendulums shown in Figure 5.24 are a familiar example from the theory of small vibrations. Two pendulums are free to swing through 360° and are weakly coupled by a spring with spring constant k. They illustrate the theorem in all its details. The kinetic energy is

$$T = \frac{1}{2}ml^2(\dot{\phi}_1^2 + \dot{\phi}_2^2) \tag{5.54}$$

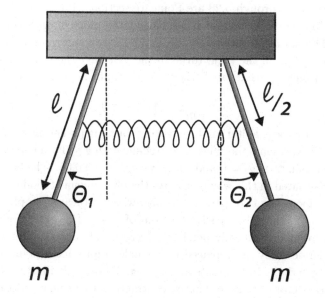

Fig. 5.24 Two coupled pendulums.

and potential energy

$$V = -mgl(\cos\theta_1 + \cos\theta_2) + \frac{1}{2}k\left(\frac{l}{2}\right)^2 (\sin\theta_1 - \sin\theta_2)^2 \qquad (5.55)$$

The equations of motion are easy to find using Lagrange's equation

$$0 = \ddot{\theta}_1 + \omega^2 \sin\theta_1 + \eta\omega^2(\sin\theta_1 - \sin\theta_2)\cos\theta_1 \qquad (5.56)$$

$$0 = \ddot{\theta}_2 + \omega^2 \sin\theta_2 + \eta\omega^2(\sin\theta_2 - \sin\theta_1)\cos\theta_2 \qquad (5.57)$$

where

$$\omega \equiv \sqrt{\frac{g}{l}} \qquad \eta \equiv \frac{kl}{4mg} \qquad (5.58)$$

In the small-angle approximation these equations can be decoupled by using a new pair of angles $\theta_\pm \equiv \theta_1 \pm \theta_2$. Then there are two normal modes with frequencies, $\omega_+ = \omega$, where the two pendulums swing in phase and the spring is not stretched, and $\omega_- = \omega\sqrt{1+2\eta}$, in which case the two pendulums swing 180° out of phase. At slightly higher energies the normal modes disappear and a new mode appears in which the two pendulums swing more or less independently. At higher energy the motion becomes wildly chaotic, but by carefully tuning the energy and coupling we can see the gradual onset of chaos exactly as predicted by the PB theorem.

Figures 5.25 through 5.28 are Poincaré sections. Each point represents the coordinates of the second pendulum at the moment when $\theta_1 = 0$ and $\dot{\theta}_1 > 0$. The initial conditions, θ_1, $\dot{\theta}_1$, and θ_2 are chosen randomly. Then $\dot{\theta}_2$ is chosen so that all the trajectories in the plot have the same energy. The energy parameter e that labels the various graphs is defined by

$$e = \frac{2}{ml^2}(T + V_g + V_c) \qquad (5.59)$$

(V_g and V_c are the gravitational and spring coupling potentials respectively.) Figure 5.25 represents the small angle regime with $\theta \leq 3°$. There are two elliptic fixed points at $\theta_2 = 0$ and roughly $d\theta_2/dt = \pm 0.05$. The fact that the loops are situated in this way illustrates the phenomena of mode locking. In the upper loops the two pendulums are swinging (nearly) in phase and in the lower loops they are (nearly) 180° out of phase. However, mode locking is a delicate affair that only persists at very small angles and energies. At higher energies they are replaced by a single elliptic fixed point at $\theta_2 = 0$ and two hyperbolic fixed points at $\theta_2 = \pm\pi$. Figure 5.26 illustrates the two kinds of trajectories. Those that are contractible to a single point represent motion within a limited range of angles. The trajectories that cannot be so

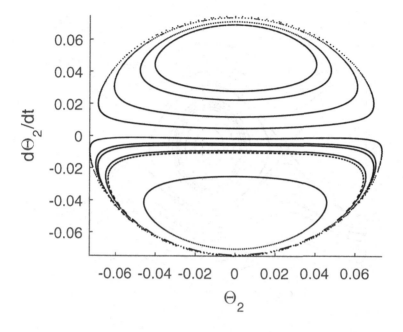

Fig. 5.25 $e = 0.0056, \eta = 0.01$.

contracted represent motion through a full 360°. As the coupling is tightened new features appear. In Figure 5.27 we see chaos developing around the initial hyperbolic fixed points. There are also "necklaces" of small loops. In terms of the Poincaré-Birkhoff theorem, these result from the sequences of elliptic and hyperbolic fixed point predicted by the theorem. Finally in Figure 5.28 chaos is spreading from the initial hyperbolic fixed points and numerous new fixed points are proliferating among the smaller loops. A further increase in e or η would cause a complete breakdown of orderly motion with one small island of sanity around the initial elliptic fixed point.

5.10   The KAM theorem: background

According to the Liouville integrability theorem from Section 3.4, a Hamiltonian H_0 with n degrees of freedom and n independent constants of motion in involution is integrable or as it is sometimes called, classically integrable or integrable in the sense of the Liouville integrability theorem. Such Hamil-

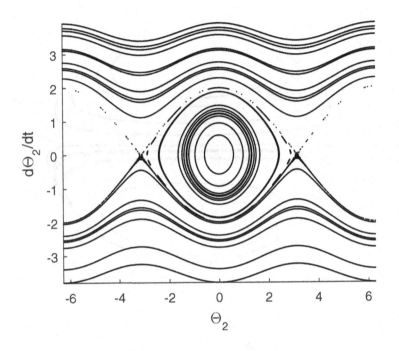

Fig. 5.26 $e = 40, \eta = 0.02$.

tonians can be expressed in terms of action-angle variables

$$\boldsymbol{\omega} \equiv (\omega_1, \omega_2, \cdots, \omega_n) \quad \boldsymbol{I} \equiv (I_1, I_2, \cdots, I_n) \quad \boldsymbol{\psi} \equiv (\psi_1, \psi_2, \cdots, \psi_n)$$

$$(5.60)$$

The frequencies can be calculated from

$$\omega = \frac{\partial H}{\partial \boldsymbol{I}} \qquad (5.61)$$

Each torus can be characterized by a unique set of frequencies $\boldsymbol{\omega}$ or alternatively by a set of action variables $\boldsymbol{I}$. Given $\boldsymbol{I}$ we could calculate $\boldsymbol{\omega}$ and vice versa. (An important caveat will appear later.) Each phase space trajectory is confined to a particular torus where it moves perpetually. We call such motion conditionally periodic.

In fact there are very few systems known that satisfy these conditions. From a historical perspective the most important system that does not is the solar system. If one neglects the gravitational attraction of the planets with one another the system is precisely integrable and the solutions can be found in any undergraduate mechanics text. But even though the interplanetary

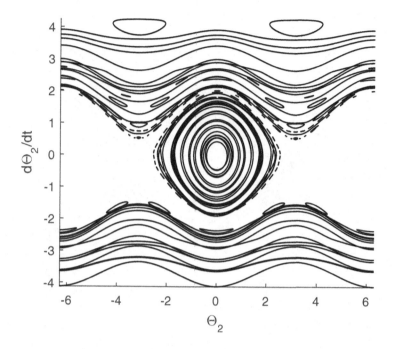

Fig. 5.27 $e = 40, \eta = 0.2$.

forces are weak, there is no obvious reason why their mutual perturbation might not lead to chaotic behavior over the history of the solar system. This problem bothered Isaac Newton who concluded that God must constantly intervene to keep things running smoothly. Laplace later proved that the solar system is stable but only in an approximation in which one ignores higher-order terms in the perturbation theory expansion.

The problem can be generalized as follows. Assume a Hamiltonian of the form

$$H(\boldsymbol{I}, \boldsymbol{\psi}, \epsilon) = H_0(\boldsymbol{I}) + \epsilon H_1(\boldsymbol{I}, \boldsymbol{\psi}) \tag{5.62}$$

H_0 is completely integrable in the sense described above. H_1 represents some perturbation, the mutual gravitational attraction of the planets for example. The ϵ is a small parameter used for book keeping. We will look for some perturbation series expansion in powers of ϵ. Henri Poincaré called equation (5.62) the "fundamental problem of dynamics" and in fact most nonlinear mechanics problems can be cast in this form. We have already used this formulation in our study of perturbation theory. We discovered

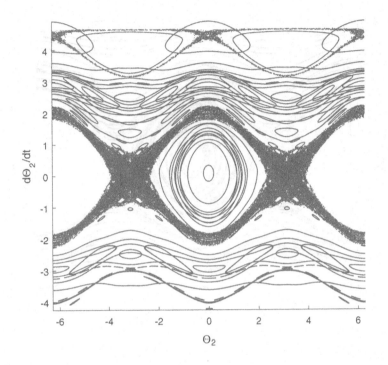

Fig. 5.28 $e = 40, \eta = 0.5$.

that there are two series that must converge (4.27) and (4.28) repeated here
for convenience.

$$\tilde{H}_1(\boldsymbol{I}, \boldsymbol{\psi}) = \sum_{\boldsymbol{k}} A_{\boldsymbol{k}}(\boldsymbol{I}) e^{i\boldsymbol{k}\cdot\boldsymbol{\psi}} \tag{5.63}$$

$$F_1(\boldsymbol{I}, \boldsymbol{\psi}) = \sum_{\boldsymbol{k}} B_{\boldsymbol{k}}(\boldsymbol{I}) e^{i\boldsymbol{k}\cdot\boldsymbol{\psi}} \tag{5.64}$$

where $\boldsymbol{k}$ is a vector of integers[3]

$$\boldsymbol{k} = k_1, \ldots, k_n$$

and

$$B_{\boldsymbol{k}} = i\frac{A_{\boldsymbol{k}}}{\boldsymbol{\omega}\cdot\boldsymbol{k}} \tag{5.65}$$

───────────────────

[3]The sum over $\boldsymbol{k}$ means the sum over all possible combinations of the n integers
$k_1, \ldots, k_n$.

The rate of decrease of the $|B_k|$ depends both on the $|A_k|$ and the denominators $|\boldsymbol{\omega} \cdot \boldsymbol{k}|$, so even if the $|A_k|$ decrease fast enough for (5.63) to converge, (5.64) will not converge if the $|\boldsymbol{\omega} \cdot \boldsymbol{k}|$ decrease too rapidly.

The situation seems hopeless. If any of the ω's yields a rational winding number, the series will blow up immediately, and if one is working with finite precision – on a computer for example – *every* number is a rational number. We came to the same conclusion in proving the Poincaré-Birkhoff theorem. Every resonant torus *i.e.* one with a rational winding number is destroyed by the slightest perturbation. And yet here we are, 4.5 billion years after the creation of the solar system. Was Newton right after all?

In 1889 King Oscar II of Norway and Sweden announced a scientific prize competition to honor his 60th birthday. There were three problems. Presumably the mathematician who came up with the most brilliant solution to one of them would win the prize. One problem was simply, prove "the stability of our planetary system." This question was included at the suggestion of the great German mathematician Karl Weierstrass. Weierstrass probably assumed that Poincaré, one of the most brilliant mathematicians of his age, would take the bait, prove the theorem, and claim the prize. Things did not work out quite that way. In 1890 Poincaré published his 270-page prize wining manuscript, "Sur le problème des trois corps et les équations de la dynamique." (On the three-body problem and the dynamic equations)[4] in which he proved, not that the solar system was stable but that it wasn't! More specifically he proved that "most"[5] Hamiltonian systems are not integrable in the classical sense. He also analyzed the heteroclinic and homoclinic tangles that we studied in the previous section. He speculated that there were no exact solutions to the n-body problem with $n \geq 3$, but was careful to note that this was a speculation only; he had not proved it.

Partly due to Poincaré's reputation and partly due to the work of his disciples, most notable the American mathematician George David Birkhoff, it came to be believed that all non-trivial dynamical systems must be unstable and chaotic. Whereas this might be bad news for us denizens of the solar system it was good news for statistical mechanics. One of the foundational principles of statistical mechanics is the so-called ergodic hypothesis.[6] Ludwig Boltzmann used it to derive the second law of thermo-

---

[4][Poincare (1890)].

[5]That innocent sounding word "most" can be made mathematically precise. The sense of it is that integrable systems are exceptional and atypical.

[6]This is discussed more fully in the Appendix.

dynamics with the following (oversimplified) argument. Consider a system of n particles in a closed volume. Phase space consists of $6n$ variables. We can imagine dividing it up into enumerable small region called microstates, each microstate representing distinct values of the $6n$ variables. Now consider macroscopic properties such as pressure, temperature, density, *etc.* and conceptually divide the space of microstates into larger regions called macrostates, each macrostate representing distinct values of the macroscopic variables. This last step is called "coarse graining." Now suppose the system starts out in one particular microstate. As it evolves it traces out a trajectory moving from one microstate to another. The crucial assumption at this point is that all the microstates have equal probability associated with them. There is no reason why trajectories might visit one microstate in preference to another. This is the ergodic hypothesis. It may or may not be true, but if it is true the second law of thermodynamics follows trivially. Some of the macrostates, those associated with equilibrium or near-equilibrium conditions, will be vastly larger than others. It is most probable that the system will find itself in the largest macrostate, and because it is so large the system will never get out. We identify the size of the macrostate with entropy according to Boltzmann's famous formula

$$S_i = k \ln W_i$$

where S_i is the entropy of the ith macrostate which contains W_i microstates. The constant k is chosen to make the units come out right. It seemed in the post-Poincaré era that the ergodic hypothesis was justified by the (supposed) fact that chaos would always "thermalize" the system so that it could visit each microstate with equal probability.

But there are two loopholes in Poincaré's conjecture, both arising from well-known results in mathematical analysis. I've said glibly that every irrational number can be approximated to arbitrary accuracy by a rational number. True – but somewhat paradoxically, the rational numbers constitute a set of measure zero within the reals. If you were to pick a number completely at random from the closed set $[0, 1]$, there is a vanishingly small probability that you would pick a rational. This can be proved with the following argument. Consider again the interval $[0, 1]$ and put all the rational numbers in one-to-one correspondence with the real integers. The sequence

$$1, \ 1/2, \ 1/3, \ 2/3, \ 1/4, \ 2/4, \ 3/4 \cdots$$

shows one way of doing this. Let the integer c_i represent the ith element in this sequence. We now "cut out" a small region of width $\epsilon/2^i$ around each rational number and sum all the areas.

$$\sum_{i=1}^{\infty} \epsilon/2^i = \epsilon$$

Where we have used the fact that this geometric series sums to one. Since ϵ can be as small as we like, we conclude that "almost"[7] every number in the set is irrational.

The second loophole concerns power series solutions of the form (5.63) and (5.64). These are Taylor series-like expansions in powers of ϵ. The solutions are in the form

$$c_0 + c_1\epsilon + c_2\epsilon^2 + c_3\epsilon^3 + \cdots$$

We call this linear convergence. There are other approximation techniques that yield quadratic convergence, *i.e.* series like

$$c_0 + c_1\epsilon + c_2\epsilon^2 + c_3\epsilon^4 + c_4\epsilon^8 + \cdots + c_n\epsilon^{2(n-1)} + \cdots$$

One such technique was invented by Isaac Newton and is variously called Newton's method, the Newton-Raphson method, or the method of tangents for finding the root of a function. It works like this: to find the value of x such that $f(x) = 0$ proceed as follows

- Choose a point x_0 on the real axis hopefully close to the exact root. Evaluate the derivative of f at this point

$$\left.\frac{df(x)}{dx}\right|_{x=x_0} = f_0'$$

- The tangent to the curve at this point can be extrapolated to the real axis. This is the first approximation to the exact root.

$$x_1 = x_0 - \frac{f_0}{f_0'}$$

- Evaluate the derivative at this new point and repeat the procedure.

$$x_2 = x_0 + x_1 - \frac{f_1}{f_1'}$$

It can be shown that the error in this procedure decreases like ϵ^{2n}. To put it another way, if ϵ_n is a measure of the error at the n step, then $\epsilon_{n+1} \approx \epsilon_n^2$. The reason for this fast convergence is that each new correction is added, not to the zeroth-order term as in the case for Taylor series-like expansions, but to the previous best guess for the value of the root. This is an idea that can be generalized in many ways. If this procedure could be adapted to canonical perturbation theory perhaps the rapidly decreasing size of the correction terms would outstrip their increase due to the small denominators.

---

[7]There's that "almost" again.

5.11   Statement of the theorem

At the 1954 International Congress of Mathematicians in Amsterdam, Andrey Nikolaevich Kolmogorov, one of the leading innovators in mathematics in the 20th century, sketched out a theorem with enormous implications; Poincaré's hypothesis was wrong and the ergodic hypothesis was questionable! For sufficiently small ϵ many of the tori with irrational winding numbers are distorted but preserved, and in the limit $\epsilon \to 0$ the measure of those that are destroyed shrinks to zero just as is the case for the rational numbers themselves. In 1963 the theorem was finally proved by his brilliant student Vladimir Arnold. (It was part of his thesis for his doctor of science degree!) A slightly different version of the theorem was independently proved and published by the German-American mathematician Jürgen Moser in 1962. Because of their contributions it is now referred to as the KAM theorem. Many people have subsequently refined, extended, and utilized it. Together their work is called KAM theory. Since Kolmogorov himself couldn't (or at least didn't) prove the theorem you may be reassured that the proof is quite non-trivial.[8] A more precise statement of the theorem follows.

Theorem 5.5. *If an unperturbed system is nondegenerate, then for sufficiently small conservative perturbations, most non-resonant tori do not vanish, but are only slightly deformed, so that in the phase space of the perturbed system there are invariant tori with phase curves winding around them conditionally periodically, with a number of independent frequencies equal to the number of degrees of freedom. The measure of the tori that are not preserved is small and goes to zero in the limit $\epsilon \to 0$.*

5.11.1   *Two conditions*

One of the conditions that must be met if the theorem is to be satisfied is that the system be nondegenerate. The sense is this: the frequencies of the unperturbed tori are found simply by $\omega = \partial H_0 / \partial I$, but we need more than this. We require that there be a one-to-one correspondence between the action variables and the frequencies so that each torus can be labeled with a unique set of frequencies and the structure of the frequency space

---

[8][Jose and Saletan (1998)] contains an outline of the proof. [Dumas (2014)] has a translation of Kolmogorov's original paper. A careful discussion is given in Appendix 8 of [Arnold (2010)] along with a bibliography of the actual proofs. Arnold modestly refers to it as "Kolmogorov's theorem."

(see next paragraph) is preserved in the space of tori. The easiest way to insure this is[9]

$$\det \left| \frac{\partial \boldsymbol{\omega}}{\partial \boldsymbol{I}} \right| = \det \left| \frac{\partial^2 H_0}{\partial^2 \boldsymbol{I}} \right| \neq 0 \qquad (5.66)$$

Note that the variables here refer to the unperturbed tori.

There is a second condition restricting the frequencies. Of course we are only considering frequencies with irrational winding numbers. Even if the frequencies are incommensurate, $|\boldsymbol{\omega} \cdot \boldsymbol{k}|$ could be arbitrarily small. The KAM theorem requires that it be bounded from below by the so-called "weak diophantine condition."[10]

$$|\boldsymbol{\omega} \cdot \boldsymbol{k}| \geq \gamma |k|^{-\kappa} \text{ for all integer } \boldsymbol{k} \qquad (5.67)$$

where $k = \sqrt{\boldsymbol{k} \cdot \boldsymbol{k}}$. γ and $\kappa > n + 1$ are positive constants.

What is the significance of this strange inequality?

- Equation (5.67) makes irrationality a quantitative concept.[11] Those irrationals that satisfy (5.67) are "more irrational" than those that don't, and the extent of their irrationality can be quantified by that value of γ for which they just do or do not satisfy the inequality. We expect that as the perturbation becomes stronger the value of γ must increase and there is some threshold perturbation at which the last surviving torus is destroyed.

- The set of numbers satisfying (5.67) is a strange Cantor-like fractal set. As the mathematicians would say, it's not "nice." It has been speculated that the bizarre nature of this set contributed to the difficulty mathematicians have had in proving the convergence of the perturbation series.

- The KAM theorem gives us no clue how to calculate the appropriate values of γ and κ or the values of ϵ for which chaos will set in. Some estimates placed the critical value of ϵ to be something around 10^{-50}! If this were true, of course, the theorem would be quite pointless. Numerical test with specific models have found critical values of ϵ as large as $\epsilon_c \approx 1$. According to [Jose and Saletan (1998)], "To our knowledge a rigorous formal estimate of a realistic critical value for ϵ remains an open question."

---

[9]Several other nondegeneracy conditions have been suggested for special classes of problems, for example Arnold's isoenergetic condition and Rüssmann's condition. See [Dumas (2014)] for further discussion.

[10]The term "Diophantine" refers to the 3rd century Hellenistic mathematician Diophantus of Alexandria who worked on problems in which only integer solutions are sought.

[11]This can also be quantified in terms of continued fraction expansions. See [Hand and Finch (1998)] or [Jose and Saletan (1998)].

5.12   Analysis

There are several important points to be made about the theorem and its consequences. In this section I will be largely paraphrasing the analysis in [Arnold (2010)].

Suppose we choose a set of frequencies ω of the unperturbed system that satisfy (5.67) above for all integer vectors k. It can be shown that for κ sufficiently large (say $\kappa > n + 1$) the measure of the set of nearby vectors ω that violate the non-resonant condition is small when γ is small.

Notice that in our perturbation theory approach in the previous chapter, the frequencies evolved with the perturbation since according to (4.2) $\dot{\psi} = \partial H(\psi, I, \epsilon)/\partial I$. The idea behind Kolmogorov's expansion is quite different. The frequencies are fixed at each order of the expansion. Each surviving torus however distorted can be labeled with the same unique set of frequencies which it inherits from the unperturbed torus. So pick out an unperturbed torus and turn on the perturbation. The unperturbed torus disappears of course, but nearby there appears another torus with the same frequencies and a slightly different set of I, *i.e.* slightly different initial conditions.

The success of Kolmogorov's perturbation scheme depends on on the rapid convergence of Newton's method, which after n iterations produces an error proportional to ϵ^{2n}. In Arnold's colorful expression this super-convergence "paralyzes" the effect of the small denominators and succeeds in carrying out an infinite series solution and proving that the series converges.

The assumption under which all this can be done is that the unperturbed Hamiltonian function $H_0(I)$ is analytic and non-degenerate and that the perturbing Hamiltonian function $\epsilon H_1(I, \psi)$ is analytic and 2π periodic in the angle variables. In some sense the size of ϵ is irrelevant. What is only important is that the perturbation be sufficiently small in some complex neighborhood of radius ρ of the real plane of the variables ψ.[12]

There is a stronger statement that can be made for systems with two degrees of freedom. Phase space is four-dimensional, but since energy is conserved there are only three independent variables. Tori are two-dimensional surfaces so they "divide" the three-dimensional space. Unstable tori that are trapped inside a stable torus can never get out and those outside can

---

[12]Kolmogorov's proof depends on continuing H and its variables into the complex plane. This makes available powerful theorems from complex analysis that are not available for real variables.

never get in. When the perturbation is small there are small isolated islands of instability trapped between stable tori. As the perturbation is increased the stable tori are overrun by regions of instability. At some critical value of ϵ the last stable torus is destroyed and the resonant or near-resonant trajectories roam chaotically over all phase space.[13] It is interesting to look back at some of our previous plots, particularly Figures 5.27 and 5.28, to see how this looks on the Poincaré section.

The question naturally arises, what happens to the unstable tori in higher-dimensional spaces? They are free to roam wandering around the stable tori, but they often do so very slowly. The first comprehensive results in this direction were laid out in the 1970's by Arnold's student N. N. Nekhoroshev. It has since turned into a huge body of results loosely referred to as "Nekhoroshev theory." A prototype Nekhoroshev-style theorem is given in [Dumas (2014)]. The gist of it is that subject to numerous technical requirements the actions $I(t)$ evolve on a time scale proportional to ϵ^{-a}, where a is "some positive constant."

5.13 Conclusion

This is the end of our story about chaos. Remember that we have only dealt with bounded, conservative systems with time-independent Hamiltonians. (Classical mechanics is a big subject.) Systems with one degree of freedom are trivial (in principle) to solve using the method of quadratures. Systems with n degrees of freedom are trivial (again in principle) if they have n constants of motion in involution. Such a system can be reduced by using action-angle variables to an ensemble of uncoupled oscillators. These systems are said to be integrable and they do not display chaos. The trouble comes when we introduce some non-integrability as a perturbation. Perturbation theory is straightforward with one degree of freedom, but with two or more degrees of freedom comes the notorious problem of small denominators.[14] Perturbation theory fails immediately for all periodic trajectories with rational winding numbers. According to the Poincaré-Birkhoff theorem, these trajectories on the Poincaré section break up into complicated whorls and tangles surrounded by regions of stability corresponding to irrational winding numbers. The notion of rational and irrational is much more subtle that you might imagine, however. Yes

---

[13]You have probably read science fiction novels with similar plots.

[14]There are other ways of doing perturbation theory in addition to the one described here. They all suffer the same problem.

there are an infinite number of rational numbers, even in the interval $[0, 1]$, but they constitute a set of measure zero, and all other numbers can be labeled by how close they are to the nearest rational number. In this sense some numbers are more irrational than others. According to the KAM theorem, as the strength of the perturbation is increased, regions in phase space break down with those with more irrational winding numbers surviving those with less. At last "Universal darkness covers all," [15] and the trajectories though deterministic show no order or pattern.

This is a summary, perhaps even a trivialization of the vast body of KAM theory. [16] I have not dealt with any of its applications nor with the fascinating business of what happens in higher-dimensional space as the tori begin to break down. You are encouraged to look at the books listed below for more detailed information.

5.14   Sources and references

Alligood, K. T., Sauer, T. S., and Yorke, J. A. (1996). *Chaos: An Introduction to Dynamical Systems*, (Springer-Verlag).

Arnold, V. I. (2010). *Mathematical Methods of Classical Mechanics, Second Edition* (Springer).

Capinski, Marek and Kopp, Ekkehard (2004). *Measure, Integral and Probability* (Springer-Verlag).

Dumas, H. Scott (2014). *The KAM Story* (World Scientific).

Goldstein, H., Poole, C., and Safko, J. (2002). *Classical Mechanics* (Addison Wesley).

Hand, L. N. and Finch, J. D. (1998). *Analytical Mechanics* (Cambridge).

Jose, J. V. and Saletan, E. J. (1998). *Classical Dynamics: A Contemporary Approach*, (Cambridge University Press).

Strogatz, S. H. (2015). *Nonlinear Dynamics and Chaos* (Westview Press).

Tabor, M. (1989). *Chaos and Integrability in Nonlinear Dynamics* (John Wiley & Sons).

Waters, Peter (1982). *Introduction to Ergodic Theory* (Springer).

Wiggins, S. (1990). *Introduction to Applied Nonlinear Dynamical Systems and Chaos* (Springer-Verlag).

---

[15] "Thy hand great Dulness! lets the curtain fall. And universal darkness covers all." Alexander Pope, The Dunciad.

[16] [Dumas (2014)] has a bibliography with about 630 entries.

Chapter 6

Computational projects

6.1 The Henon-Heiles Hamiltonian

To model the motion of stars in our galaxy, in the 1960's Henon and Heiles proposed a Hamiltonian with two degrees of freedom x and y. The Hamiltonian is

$$H = \frac{1}{2}(p_x^2 + p_y^2) + V(x, y)$$

The potential energy is

$$V(x, y) = \frac{1}{2}\left(x^2 + y^2 + 2x^2y - \frac{2y^3}{3}\right)$$

(1) Prove that there are four fixed points, which depend only on x and y and plot them on the x-y plane. Sketch some approximate equipotential curves in the x-y plane.

(2) One of these points is stable; the other three unstable. Which one is stable? What is $V(x, y)$ close to the three unstable equilibrium points?

(3) Prove that the motion is bounded if the total energy $E < 1/6$ and unbounded if $E > 1/6$. It will turn out that chaos sets in when E is less than but close to $1/6$.

(4) Write a program to integrate the equations of motion. Find the Poincaré section for $x = 0$, $\dot{x} > 0$ and also plot the boundary which occurs at $\dot{x} = 0$. Find the threshold for chaos and observe whether the chaotic trajectories are local or global as the energy is raised.

(5) The appearance of chaos proves that the Hamiltonian is not integrable. Now change the sign of the last term (proportional to y^3). Again look at the Poincaré section. Verify that this seemingly small change leads to complete integrability at all energies where the motion is bounded. [Goldstein *et al.* (2002)] have a long section on the H-H Hamiltonian.

Use the program to investigate some of the points raised in these sections that seem interesting to you.

6.2   The orbit of Mercury

The precession of the perihelion of Mercury was one of the first tests of general relativity. There are many more tests available now, but the precession has become a standard textbook example of nonlinear dynamics. The equation of motion comes from requiring the planet to follow a geodesic path in space-time curved by the gravitational potential. This small correction causes the Kepler ellipse to gradually rotate or precess its spatial orientation in the orbital plane. The radial equation in standard notation is

$$\frac{d^2u}{d\phi^2} + u - \frac{1}{p} = 3mu^2$$

The dependent variable is $u = 1/r$, and ϕ is the angular position of the planet. It is simply proportional to time, since the angular momentum is conserved.

There are two characteristic distances in this equation, one arising from classical physics and one from general relativity. The classical distance is p, the radius of the orbit one would calculate assuming that it was a perfect circle. The other distance is the mass of the sun, $m \equiv GM_\odot/c^2 = 1.477$ km.

This is a good equation on which to practice the techniques we have covered for 2-d systems.

(1) Show that one can introduce a dimensionless variable X so that the equation becomes

$$\ddot{X} + X - 1 = 3\beta^2 X^2$$

where $\beta = v/c$ and the dots indicate differentiation with respect to ϕ. (Ignore the difference between mass and reduced mass, *i.e.* assume that the mass of the sun is very much larger than the mass of the planet.)
(2) Construct a 2-d state space and write the equations in our standard form.
(3) Is there any dissipation in this problem? Give mathematical reason and a physical reason for your answer.
(4) Find the two fixed points. Use β as your control parameter. Describe the behavior of the fixed points as functions of β. Might there be some

critical value of β at which everything changes? (This sounds like a question with a "yes" answer.) Describe what happens at this point.

(5) What does a planet actually do at or in the vicinity of the fixed points you have described?

(6) For the planet Mercury, $\beta = 1.6 \times 10^{-4}$. What is its precession rate?

(7) Write a computer program to integrate the equations of motion and illustrate the points you have made above. You should include a polar plot of X vs. ϕ and a plot of X vs. $\dot{X}$.

6.3 The standard map

The standard map $(\theta_n, I_n) \to (\theta_{n+1}, I_{n+1})$ is given by the equations[1]

$$I_{n+1} = I_n + k \sin \theta_n$$

$$\theta_{n+1} = \theta_n + I_{n+1}$$

(Both I and θ are taken modulo 2π.) The lines of constant action may be plotted on an I versus θ graph. These lines are the KAM tori, with winding numbers that depend on the action. When a nonlinear perturbation is added by making $k \neq 0$, the tori are distorted, and as predicted by the KAM theorem, the ones with rational or nearly rational winding numbers are broken. The winding number is the average increase in θ

$$\Omega \equiv \lim_{n \to \infty} \frac{\theta_n - \theta_1}{n}$$

(1) Prove that the Jacobian of the map is 1 and thus the map preserves area. When $k = 0$, for what values of I is Ω a rational number, giving periodic orbits?

(2) Increase k so that $k = 0.6$. Start the map off at a number of I values. You can choose $\theta_1 = 0$ and make at least several hundred points for each initial value. Observe what happens to tori with both rational and irrational winding numbers. In particular, discuss what happens in the region $I = \pi$ and near other places where the winding number is irrational.

(3) Keep on increasing k. The last KAM torus disappears for $k > 0.9716\ldots$ The winding number of this torus is the golden mean, $\Omega = (\sqrt{5} - 1)/2$. The last torus disappears, there is no further barrier to global chaos. Demonstrate these effects with suitable 2-d plots of I versus θ.

---

[1]From [Hand and Finch (1998)], Chapter 11, page 476.

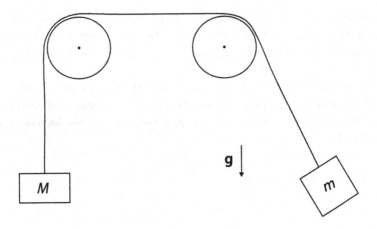

Fig. 6.1   The swinging Atwood's machine.

6.4   The swinging Atwood's machine

Figure 6.1 shows the famous swinging Atwood's machine. The original article on which this is based is [Tufillaro *et al.*(1984)]. You can get references to subsequent articles by just going to Wikipedia and typing in "Swinging Atwood's Machine." A string is hung over two pulleys. On one end is a mass M which does not swing. On the other end is a mass m which is free to swing in a vertical plane. There is a uniform gravitational field. The string and pulleys have no mass and there is no friction. Also ignore the size of the pulleys. You may assume they are points so that the length of the pendulum remains constant.

(1) Construct the Lagrangian and write down the Euler-Lagrange equations of motion. (For heaven's sake, don't try to solve them!) These two equations have simple physical interpretations. What are they?
(2) What is a cyclic (or ignorable) coordinate? Are there any in this problem?
(3) Find the canonical momenta and write down the Hamiltonian.
(4) What are the constants of motion? How many are there in this problem? How do you know?

(5) Try to find Hamilton's characteristic function. I predict that you will fail. What goes wrong?

(6) Now turn off gravity, *i.e.* set $g = 0$. What are the constants of motion?

(7) Now it should be easy to find the characteristic function; but don't do any integrals just yet.

(8) Find the equations of motion. The following integral should be useful.

$$\int \frac{dr}{\sqrt{a - b^2/r^2}} = \frac{1}{a}\sqrt{ar^2 - b^2}$$

(9) Could you solve this problem using action-angle variables? What might go wrong?

Redo the problem without the $g = 0$ approximation. Write your own program to solve the equations of motion and plot the results. Try the following:

(1) See if you can reproduce your own results for the $g = 0$ case.

(2) Reproduce a few of the published plots.

(3) It is known that the system is integrable if $M/n = 3$. How could you tell this by looking at your plots?

(4) Under some circumstances the motion is known to be chaotic. Can you find such regimes? How can you tell if it's chaotic?

Appendix A

Measure theory and the ergodic hypothesis

What are we to make of Boltzmann's ergodic hypothesis in the light of the KAM theory? If most orbits simply wind their way around invariant tori, how can they visit all the microstates with equal probability? Boltzmann's definition of entropy suffers from a related problem. Microstates and macrostates represent "distinct" regions of microscopic and macroscopic variables. But if physical properties vary continuously, what one regards as distinct is purely arbitrary. The alternate hypothesis that every trajectory visits every point in phase space is mathematically absurd. Boltzman proposed another approach, if not to calculate S, at least to define more clearly what is meant by the ergodic hypothesis. Suppose we deal not with distinct states but with averages. Let x be a state at time $t = 0$ and $\varphi_t(x)$ represent the state at time t. When we measure some macroscopic quantity $f(x)$ we never measure its instantaneous value but its value averaged over time T.

$$\overline{f}^* = \frac{1}{T} \int_0^T f(\varphi_t(x))dt \tag{A.1}$$

The implication is that the integral exists and the average $\overline{f}^*$ is independent of T so long as T is large compared with some microscopic time scale. The trouble is that if φ_t really takes in the detailed microscopic behavior of the system, we have no way of calculating it. So if we can't average over time, perhaps we can average over space.

$$\overline{f} = \frac{\int_M f(x)dx}{\int_M dx} \tag{A.2}$$

Here the integral ranges throughout phase space on the constant-energy manifold M. Boltzmann's idea is that the orbit $\varphi_t(x)$ ranges through phase space in such a way that the averages in (A.1) and (A.2) are equal. This is the ergodic hypothesis. There's a slogan that goes with this definition, time averages equal space averages.

There are many things about these integrals that are vague and/or poorly defined. A rigorous formulation of what is really a very good idea had to await the development of measure theory in the 20th century. This is no place to review an entire branch of mathematics, but a few comments might help.[1] A measure on a set is a systematic way to assign numbers to each suitable subset of that set, numbers that can intuitively be understood as generalizations of familiar quantities such as length and probability. Measure theory is a way of doing mathematics with sets and functions for which the usual rules of algebra and calculus don't apply. To this end measure theory is always introduced using the machinery of set theory. Let Ω be a set, *e.g.* all the numbers on the real line $\mathcal{R}$. We start with the power set, usually written 2^Ω. This is the set of all possible subsets of Ω including the empty set $\emptyset$ and the set Ω itself. Our goal is to assign to each of these subsets a number $\rho \geq 0$ called the *measure* of the subset. If this is to have any connection with our intuitive notion of length it must have two commonsense properties.

$$\rho(\emptyset) = 0 \tag{A.3}$$

If $E_1, E_2, \ldots$ are pairwise disjoint subsets, *i.e.* if

$$E_i \bigcap E_j = \emptyset \text{ for } i \neq j \tag{A.4}$$

then

$$\rho\left(\bigcup_{i=1}^\infty E_i\right) = \sum_{i=1}^\infty \rho(E_i) \tag{A.5}$$

a property called countable additivity. Unfortunately, there are many sets for which this cannot be done consistently. In order to make this work we must create a new set $\mathcal{A}$ with some additional conditions. In order to be measurable $\mathcal{A}$ must satisfy four additional requirements

$$\mathcal{A} \in \Omega \tag{A.6}$$

$$\text{if } E \in \mathcal{A} \text{ then } E^c \in \mathcal{A} \tag{A.7}$$

Here E^c is the complement of E, *i.e.* everything in Ω that is not in E.

$$\text{if } E_i \in \mathcal{A} \text{ then } \bigcup_{i=0}^\infty E_i \in \mathcal{A} \tag{A.8}$$

---

[1]See [Capinski (2004)] for an accessible introduction.

Finally we require countable additivity from equation (A.5) above. Sets satisfying these conditions constitute a σ-algebra or σ-field. The triple $(\Omega, \mathcal{A}, \rho)$ is called a *measure space*.

So far I have not said anything about *how* we assign numbers to subsets. There are many ways of course. The most commonly used is *Lebesgue measure* which we have already encountered in Section 5.10. We set out to find the "length" of the subset containing only the rational numbers in the interval $[0, 1]$. Conventional mathematics has no way of making sense of the question, but we came up with the answer zero. We did this by blocking out a finite interval around each rational number. These small intervals were amenable to conventional math. We added up their length and showed that in the limit $\epsilon \to 0$, the total length was smaller than any positive number. Such sets are said to have measure zero; they are *null* sets. That was easy; there are many ways of choosing the intervals that would have given the same answer. But what if the set isn't null, what then? Suppose we are given a set E (open or closed, it doesn't matter) of real numbers. We choose a set of countably infinite open disjoint intervals that *cover* each element in E. We calculate the total length by adding up the lengths of the individual intervals. This length may easily overestimate E because E is a subset of the union of the intervals, and so the intervals may include points which are not in E. The Lebesgue measure is the greatest lower bound (infimum) of the sum of the lengths among all possible such sets of intervals. Intuitively, it is the total length of that set that fits E most tightly. More formally, given I_k a sequence of disjoint intervals with

$$E \subseteq \bigcup_{k=1}^{\infty} I_k \tag{A.9}$$

$$\rho(E) = \inf\{\sum_{k=1}^{\infty} l(I_k)\} \tag{A.10}$$

where $l(I_k)$ is the length of the kth interval.

In the previous examples the set Ω consisted of points along the real line $\mathcal{R}$, but the integrals in (A.1) and (A.2) range over $6n$-dimensional phase space. Fortunately all the ideas of Lebesgue measure carry over to $\mathcal{R}^N$ with very little modification, except now *measure* refers to the measure-theory generalization of N-dimensional *volume*. The phase space volume is defined by

$$\sum_{i=1}^{3N} dq_i dp_i \tag{A.11}$$

This is the volume we are "measuring," but this is exactly the volume which, according to Liouville's theorem, is conserved under Hamiltonian flow. For this reason (A.11) is sometimes called the Liouville measure. So with φ_t as defined above, and ρ the Liouville measure, ρ is invariant with respect to φ_t. In other words, ρ is an *invariant measure*.

Measure theory is also the foundation of the modern axiomatic formulation of probability and statistics. The first step is to define a probability measure. If the measure of the entire set Ω is finite it can be used to normalize the Lebesgue measure, so that $P(E) = \rho(E)/\rho(\Omega)$. Obviously $P(\Omega) = 1$, and P has all the common sense properties that we associate with probabilities. It's called the *probability measure* or simply the *probability*. The triple $(\Omega, \mathcal{A}, P)$ is called a *probability space*. In this context the elements of $\mathcal{A}$ are often referred to as *events*.

Just as measure was introduced to deal with sets that are not amenable to ordinary mathematics, so it is necessary to generalize the notion of integration to deal with functions that are not sufficiently well-behaved to do the standard Riemannian integration. In the Riemann integral we split the interval $I = [a, b]$ over which we are integrating into smaller intervals I_n. The easiest way to do this is to divide I into N equal intervals. To integrate over the function $f(x)$ we make approximating sums like

$$\sum_{n=1}^{N} I_n c_n$$

where c_n is some number between $f(x_{n-1})$ and $f(x_n)$. In the limit $N \to \infty$ the sum converges to $\int_b^a f(x)dx$, but only if f is sufficiently regular.

The alternative is called *Lebesgue integration*. The basic strategy of taking the limit of sums over small rectangles is the same. The difference is that rather than slicing the function vertically, we slice it horizontally. More exactly, we divide up the range (rather than the domain) of the function into small intervals I_n and form the sum

$$\sum_{n=1}^{N} c_n \rho(f^{-1}(I_n)) \tag{A.12}$$

where c_n is some number between I_{n-1} and I_n. Do you see the power of this? No matter how bizarre the function may be, its inverse $f^{-1}(x)$ lies along the real axis, and even if this projection is discontinuous, it has only to be *measurable* for the technique to work. In fact if $f^{-1}(x)$ is measurable in some interval we say the function is *Lebesgue-measurable* over that interval.

It requires considerable ingenuity to come up with a function that fails this criterion.

We now have all the machinery ready to give a rigorous definition of ergodicity. We form the probability space (M, Σ, P) where M is the energy conserving manifold that appears in (A.1) and (A.2), Σ the σ-algebra of measurable subsets, and P is the probability measure. Let T be any measure-preserving transformation. We say that T is ergodic with respect to P (or vice versa) if one of the following equivalent statements is true.[2]

$$\text{For every } E \in \Sigma \text{ with } T^{-1}(E) = E \text{ either } P(E) = 0 \text{ or } P(E) = 1 \quad (\text{A}.13)$$

$$\text{For every } E \in \Sigma \text{ we have } P\left(\bigcup_{n=1}^{\infty} T^{-n}E\right) = 1 \quad (\text{A}.14)$$

For every two sets E and H there exists an $n > 0$ such that

$$P\left((T^{-n}E)\bigcap H\right) > 0 \quad (\text{A}.15)$$

The sense of these definitions is clear enough. If any subset is mapped back onto itself by T the system is not ergodic. If the system is ergodic then $T^{-n}E$ sweeps out all of M. In the course of doing so it intersects all other subsets. This finally is a rigorous version of Boltzmann's idea of the orbit that moves through all microstates with equal probability. We should note however that null sets have zero measure, so in principle there may be null sets that don't satisfy the above requirements even in an ergodic system.

This brings us back to the integrals in (A.1) and (A.2). The first rigorous version of the equality of the two was proved by G. D. Birkhoff in 1931.[3] Modern texts refer to it simply as "the ergodic theorem." Given a probability space (M, Σ, P) as described above, then the following statements can be proved.[4]

Theorem A.1. *Given any Lebesgue-measurable function $f(x)$ representing some observable quantity evaluated when phase space is in its initial state x, the sum*

$$\frac{1}{n} \sum_{i=1}^{n-1} f(T^i(x)) \quad (\text{A}.16)$$

---

[2]See [Waters (1982)] for proofs and numerous other definitions.
[3][Birkhoff (1931)].
[4]Waters *ibid.*

converges to a function $f^*(x)$. This is the measure-theory version of the time integral integral (A.1).

Theorem A.2. *f^* is in fact constant. It's a number.*

Theorem A.3.

$$f^* = \int f \, dP \qquad (A.17)$$

The integral ranges over all phase space at one instant in time.

There you have it; time averages equal space averages. (With the usual caveat that it might not apply to null sets.)

This is all mathematics. Start with an abstract definition and spend the rest of your career proving theorems. But do real dynamic systems behave like this? In a word – some do, some don't. This is a fascinating and active field of research for experimenters and computer modelers. Because of the KAM theorem it was expected and later proved that generic smooth Hamiltonian systems are not ergodic.[5] Other dynamical systems show a wide variety ergodic and non-ergodic behavior.

---

[5]See [Markus and Meyer (1970)] and [Markus and Meyer (1974)]. These are very technical papers, but the idea is simple, orbits around elliptical fixed points block the spread of chaos as we have seen in several of our numerical models.

Bibliography

Alligood, K. T., Sauer, T. S., and Yorke, J. A. (1996). *Chaos: An Introduction to Dynamical Systems*, (Springer-Verlag).

Arnold, V. I. (2010). *Mathematical Methods of Classical Mechanics, Second Edition* (Springer).

Birkhoff, G. D. (1931). Proof of the ergodic theorem *Proc. Nat. Acad. Sci.* **17** pp. 656–660.

Capinski, M. and Kopp, E. (2004). *Measure, Integral and Probability* (Springer-Verlag).

Carroll, S. (2010). *From Eternity To Here* (Penguin Group).

Dumas, H. S. (2014). *The KAM Story* (World Scientific).

Goldstein, H., Poole, C., and Safko, J. (2002). *Classical Mechanics* (Addison Wesley).

Hand, L. N. and Finch, J. D. (1998). *Analytical Mechanics* (Cambridge).

Jose, J. V. and Saletan, E. J. (1998). *Classical Dynamics: A Contemporary Approach*, (Cambridge University Press).

Markus, L. and Meyer, K. R. (1970). Generic Hamiltonian systems are not ergodic, in *Qualitative Methods of the Theory of Nonlinear Oscillations*, Tome 2 *Proceedings of the 5th International Conference on Nonlinear Oscillations, Aug. 25–Sept. 4, 1969* (Institute of Mathematics, Ukrainian Academy of Sciences, Kiev) pp. 311–332.

Markus, L. and Meyer, K. R. (1974). Generic Hamiltonian systems are neither integrable or ergodic, *Menoirs Amer. Math. Soc* No. 144, (Amer. Math Society).

Matzner, R. A., and Shepley, L. C. (1991). *Classical Mechanics* (Prentice-Hall).

Poincare, H. (1890). *Acta Math.* **13**, pp. 1–270.

Strogatz, S. H. (2015). *Nonlinear Dynamics and Chaos* (Westview Press).

Tabor, M. (1989). *Chaos and Integrability in Nonlinear Dynamics* (John Wiley & Sons).

Taylor, J. R. (2005). *Classical Mechanics* (University Science Books).

Tufillaro, N., Abbott, T. A. and Griffiths, D. J. (1984). Swinging Atwood's Machine *American Journal of Physics* **52** pp. 903.

Scheck, F. (1994). *Mechanics* (Springer-Verlag).

Waters, P. (1982). *Introduction to Ergodic Theory* (Springer).

Wiggins, S. (1990). *Introduction to Applied Nonlinear Dynamical Systems and Chaos* (Springer-Verlag).

Index